Rowell
Home Inspection
770-645-2855
2882

The Home Inspector's Guide to Preventative Maintenance

Finding Problems Before They're Problems!

By Stephen Showalter, RRI - *Professional Home Inspector*

LIBRARY OF CONGRESS CATALOGING IN PUBLICATION DATA

Showalter, Stephen.
The Home Inspector's Guide to Preventative Maintenance/by Stephen Showalter.
p. cm.
ISBN 0-9660775-0-4
1. Home Inspection 2. Home Maintenance
I. Title

97-094646
1998 CIP

First Edition
1 2 3 4 5 6 7 8 9 10

Design, Layout & Cover Photo by Amy Vansant
Photographs and Graphics by Stephen Showalter

http://www.hurricanepress.com
http://www.buildingspecs.com

Warning to Readers

Working on houses and with tools can be dangerous, and participants may potentially suffer bodily injury or death. The reader assumes all risks that may result from participating in any of the described activities. The authors of this book cannot be held responsible for any damage to buildings or their components as a result of following suggestions in this book.

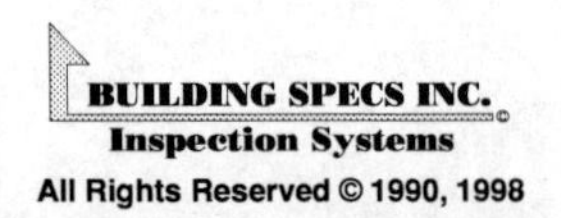

Table of Contents

Section One: The Big Picture	**2**
Routine Maintenance	3
Master Checklist of Home Inspection	4
The 3 Rule Approach to Home Inspection	6
Constants: Timing & Age	8
Childproofing Checklist	11
Section Two: Building Blocks	**12**
Structural Checks	13
Foundations	24
Basements & Crawlspaces	33
Waterproofing Recommendations	43
Section Three: Outside	**45**
Grounds	46
Exterior	52
Decks & Porches	64
Roofs	70
Fire Retardant Treated Wood (FRT)	88
Attics	90
Section Four: Plumbing	**97**
Supply & Waste	98
Winterizing	105
Bathrooms	108
Section Five: Heating and Cooling	**114**
Heating	115
Cooling	128
Chimneys & Burning Systems	131
Section Six: Electric Systems	**138**
Electrical Overview	139
Appliances	147
Section Seven: Interior	**150**
Room Surveys	151
Section Eight: Environmental Concerns	**162**
Asbestos	163
Buried Oil Tanks	165
Carbon Monoxide	168
Electromagnetic Fields	170
Lead	171
Radon	174
Septic Concerns	177
Water Contaminates	181
Inspection Summary	**187**
Inspection Graph	**188**
Glossary	**189**

Section One:

The Big Picture

Routine Maintenance
✓ Checklist

Routine Maintenance is the most important thing you can do for your home. The person who came up with the proverb "an ounce of prevention is worth a pound of cure" probably did so after being buried by home repair bills which could have been avoided. Whether you are hoping to make a profit from the sale of your home, avoid repair bills, make your house last, or just like the security of a sound property, home maintenance is the key to your long term "home happiness."

And don't be fooled by the term "home maintenance" --- keeping your property up to shape involves much more than just the floors and walls of your house. Because everything on your property is related in some way, you have to take care of your home's various components *and* the grounds which surround it.

The trick to good home maintenance is to avoid only thinking about it once every five years. Effective home maintenance is a year-round project, although the frequency may vary with region, severe weather, usage, materials, design, and other factors. Many homes are considered "maintenance free," such as a brick or vinyl sided houses. This usually means that only minor maintenance is expected, such as outside caulking. Higher maintenance buildings include homes with all wood exteriors. Homes with HVAC filters may require more frequent cleaning if there are pets in a building, and an ocean front house may require more frequent maintenance to its siding, roof or deck. But regardless of how high maintenance your home is, time spent keeping your home in shape will be returned to you in saved repair bills, return on investment, and the avoidance of sheer frustration.

Use the following checklist to prompt you as you look for symptoms; visual signs or clues could indicate a deeper or hidden problem. For example, a water leak at a block foundation wall could signify a cracked wall. A masonry sealer on an interior wall could indicate an attempt to prevent water from penetrating. You have to be part detective; look at all of the options and use the process of elimination.

And while we don't know where you live or what sort of home you own, the following list describes some of the average building's components which need to be maintained. Definitions of unfamiliar terms can be found in the back of the book or in the chapter margins.

Following is a master checklist of potential trouble spots in any home. Each of these topics will be discussed in detail in the proceeding chapters.

MASTER CHECKLIST OF HOME INSPECTION

GROUNDS

✓ Divert ground or runoff water away from the house. Look for problems with:
- a.) Sump pumps
- b.) Inadequate or negative drainage
- c.) Downspout and gutter water control

✓ Trim bushes, trees, and roots back from siding, foundation, over chimneys *(fire hazard)* and roof, to prevent damage.

✓ Inspect crawlspace for moisture problem.

✓ Keep mulch and soil from contact with siding or house framing. Prevent mounding of the soil or mulch against the house, and maintain a minimum of several inches from the bottom of the siding. When soil or mulch comes in close contact with the siding, the chances for termites and/or water damage are greatly increased.

✓ Asphalt driveways need to be sealed on a regular basis. This will prevent water from entering the cracks, freezing, and causing deterioration of the driveway.

EXTERIOR

✓ Keep gutters and downspouts clean.

✓ Caulk around exterior windows, doors, wood trim and any areas susceptible to water infiltration. Caulk seams or joints where trim is over or butting against another section of wood. Caulking in many cases is more important than paint. If water is allowed to get behind,or to wick into the end-grain, water damage or failing paint can occur. For example:
- a.) A garage door panel where the panels are seated in a rail. Water can penetrate into the rail and deteriorate both panel and rail. If the seam is caulked, the infiltration is eliminated.
- b.) Plywood panels below windows with a trim detail nailed on the face are susceptible to water damage. Caulk around the trim, as well as where water can infiltrate from the edges of the plywood.

✓ Seal wall penetrations, plumbing, electrical, etc.

✓ Paint any exterior wood.

✓ Overall, inspect for signs of rot, failing paint, mildew, fungus, water stains, or cracked caulking.

ROOF

✓ Keep debris cleared from roof, valleys, and from collecting above or behind skylights and chimneys.

✓ Inspect the condition of roofing material and all flashings.

✓ Check the edge of roof sheathing for rot due to lack of a drip edge.

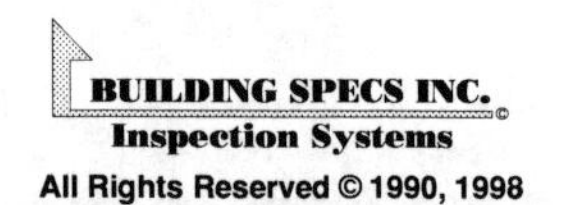

✓ Inspect around all plumbing vent flashings for deterioration.

HEATING AND COOLING

✓ Clean or replace HVAC filter.
✓ Peroxide in the condensate tray, drain line and pump can help prevent bacteria and algae from thriving (approximately a cap full in each section.)
✓ Routinely clean and service the humidifier.
✓ Annually service and clean any component that burns fuel.
  a.) Oil, natural gas or propane fired forced air furnace.
  b.) Oil, natural gas or propane fired hydronic system .
  c.) Natural gas or propane stove and/or oven.
  d.) Woodstove.
  e.) Fireplace.
  f.) Chimneys, flues, and chimney caps.

PLUMBING

✓ Routinely inspect all waste lines, traps, and supplies.
✓ Drain the hot water heater (turn off first) and refill.
✓ Fill any floor drains or unused sinks, tubs, etc. with water. This will fill the trap and prevent sewer gas from entering the house.
✓ Caulk around tub and shower fixtures to prevent water damage to the wall.
✓ Check sump pump for proper operation. Peroxide can help keep bacteria from thriving.

ELECTRICAL

✓ Test all ground fault circuit interrupters (GFCI).
✓ Test smoke and carbon monoxide detectors (with smoke).
✓ Look for any burned receptacles.

ATTIC

✓ Look for active leaks, especially around chimneys, valleys, skylights and other roof penetrations.
✓ Look at the condition of ventilation screens and for any animal nesting.
✓ Verify that power fan is operational.

DOORS AND WINDOWS

✓ Repair broken seals in insulated glass panes (trapped moisture).
✓ Replace failing weather stripping.
✓ Replace worn door sweeps.
✓ Repair condensation damage.
✓ Replace cracked glass.

The 3 Rule Approach To Home Inspection

When conducting an inspection of your home, it is a bad habit to assume the condition or status of anything, or to become focused on small objects and completely overlook the big picture. Of course, it is just as easy to bypass the smaller things by running through the house scanning for the big items. Remember, you are a detective looking for clues and symptoms; little signs can be very helpful.

If you stick to the following three basic rules, you should be able to give your home a thorough inspection, and feel safe in the knowledge that you have solved The Case of The Suspicious Home Condition.

RULE ONE

Stand back and look at the big picture. For example: an outside wall. Stand back and look straight on, let your eyes follow the wall's lines up, down and across. Look for any abnormalities, heavy shadowing, bulges, etc. Then stand at one end of the wall and sight up and across. This angle will show different rolls, bulges and even signs of sagging (which could indicate settling).

RULE TWO

Get an up-close look. For example: an outside wall. Let your eyes travel up and down each section of siding and look for irregularities such as dark shadows which with vinyl could indicate the siding not interlocking. Look for damaged sections, such as cracking or holes.

RULE THREE

Touch whatever and wherever possible. For example: an outside wall. By touching the siding you can feel if is properly interlocked, inadequately nailed or extremely loose. By wiping your hand along it you can see if is chalking.

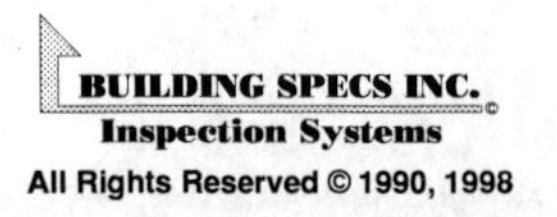

Here are two more examples to ensure you understand the basics of the 3-Rule Approach:

Example: Roof

1. Look at the shingles and roof shape from the ground; use binoculars if necessary. Get on the roof if possible, and look for overall sagging, bulging, or major abnormalities.

2. Get at one end of the shingle run and sight for overall straight runs. Climb to the peak and sight down the key ways and look for over all straightness. This will give you a quick reference of the workmanship quality or competency. Look at the valleys for tar, which could indicate an older leak.

3. See if the shingles are properly nailed by lifting the tabs if possible. Look at the plumbing vents to see if they are dry rotted.

Example: Water Heater

1. Look at the kind of system electric, gas solar etc. Check its general condition, age, location and set up.

2. Trace the plumbing for proper connections; look for corrosion at the connections, proper relief valve and direction, and signs of rust at the bottom.

3. Try to trace rust to its origin. Open the access panels, look for corrosion at the element, look for rust inside the jacket and on the tank when insulation allows, and check the thermostat setting and the water temperature.

Getting the idea? This system can be applied to everything in an inspection including heat pumps, drywall, doors, rafters and crawlspaces. Practice this method and you will become proficient at it, and be less likely to overlook something. You will also find that your inspections get easier every time you do one.

Constants:
Timing and Age

Regardless of what sort of house you have, there are two factors affecting home inspections which remain constant from property to property: the timing of your inspection, and the age of the home. This chapter will provide an overall snapshot of the various factors that could, or *should*, affect your maintenance inspection.

TIMING

Victims of Mother Nature

Temperature and weather can play a big factor in your inspection. For example, a drought may not be the optimal weather condition in which to discover a potential or active water problem in a basement. After all, it is hard to determine if a glass has a leak when it is empty. However, if it has been raining for a week and the basement is bone dry, there is a good chance that the basement is safe from active flooding. Keep the weather in mind and plan your inspections accordingly.

AGE

Historical or Hysterical?

To determine the age of the house, start by checking the title, asking the former owners or querying the real estate agent who found it for you. Also, check to see if there is a date on the electrical panel inspection, a date stamped under the toilet tank lid, or a date stamped on a brick on the house. Of course, these dates may simply indicate a remodeling date, but they offer some clues.

Born Yesterday

With new homes, expectations of a finished product with little or no problems are justified. Minor items that may need repair commonly found in a new home include:

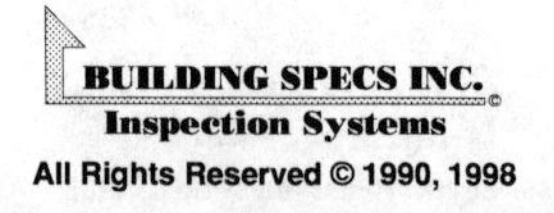

- Items found on a routine punch list
- Drywall point up
- Nail pops
- Paint touch up
- Scratches in finished wood (i.e. kitchen cabinets)
- Missing miscellaneous hardware
- Binding doors
- Cracked window panes
- Dirty HVAC vents & filter
- Improper grading

These items are usually repaired by the builder, though some builders may request that you wait for the one year walk through for drywall point up. Minor settling, tape and nail pops, and other drywall imperfections are normal. On the other hand, a new house should not show any signs of foundation settling, water infiltration, severe drywall settling, severely damaged materials (i.e. siding, roofing, trim, etc.), erosion, improperly functioning appliances or mechanical components. Many times, the builder should assume all liabilities and warranties for the house and its components. A recently renovated building may be inspected for traits found in a newer home.

> *Definitions...*
>
> **Punch List:**
> Itemized list of repairs or incomplete items for new home construction.

The Early Years

A house that is 2-10 years old may begin to show minor wear and settling. Most foundation settling will usually have occurred by now. However, if a drainage problem is left unresolved, further settling may be a result. By now some maintenance will be required, such as caulking, painting and annual mechanical maintenance. The house should be structurally and mechanically sound. These are some normal conditions expected in a house of this age:

- Drywall point up
- Nail pops
- Minor settling cracks
- Binding doors
- Paint touch up
- Improper grading

Any severe structural problems may be covered under the home or builder's warranty.

The Terrible Teens

A house that is 11-20 years old may begin to show more wear and settling. There may be a need to repair or replace some of the components, such as those affected by minor rot, peeling basement sealant, failing appliances, HVAC, shingles, siding, caulking, cosmetic surfaces, minor plumbing, concrete chimney caps, and other items. If the appliances are original, they may be nearing the end of their life. A shingled roof's normal life span is 15-25 years, though this may vary. The house should be in sound structural and electrical condition at this age.

Middle Age

As a building ages, it is normal to expect settling in areas such as the foundation, floors, walls, roof, ceilings, and other areas. Some inherent conditions due to age are normal. You should expect to upgrade some components even before their life expectancy is reached, and even the best cared for house will need some repairs. Lead paint and asbestos may be present in buildings of this age and may need to be removed. Owners of a middle-aged house just need to understand that this is not a new building, and the honeymoon is over.

Historic Buildings

The owners of an historic building should be fully aware that their homes could suffer severe settling or have outdated building techniques and components. Mortar may be failing and fireplaces may not be safe to operate. Settling, plaster failing, binding doors, inadequate electrical and heating components, and inadequate R-value for insulation and windows, are common with buildings of this age.

Summary

All of the above examples are broad generalizations chosen to provide an idea of acceptable conditions for variously aged buildings. Some older homes may be in pristine condition, while newer structures may age poorly due to various factors. The sole purpose of this synopsis is to underscore that signs of aging in buildings are expected. Much of what you discover will not necessarily be the catalyst for an immediate repair.

Definitions...

R-Value: A rating assigned to a building product and its ability to slow heat loss, such as insulation, doors and windows.

CHILDPROOFING CHECKLIST

Quick Steps You Can Take To Make Homes Safer

THINGS TO WATCH FOR	PREVENTATIVE STEPS
Are There Window Guards	☐ To Prevent Windows From Fully Opening
Cords For Blinds & Draperies	☐ Tie Them Up Out Of Reach
Electrical Cords	☐ Move Them Out Of Reach
Open Electrical Outlets	☐ Use Cover Plates
Loose Bookshelves That Can Topple	☐ Fasten To Wall
Unstable Furniture	☐ Stabilize Or Remove
Open Dresser Drawers	☐ Keep Closed So Baby Can't Climb
Fireplace Screens	☐ Use & Close So Baby Can't Get Burned
Tablecloths (That Can Be Pulled Down)	☐ Use Short Table Cloths Or None
Sharp Edges (Corners)	☐ Use Soft Guards
Scatter Rugs (That Can Slip)	☐ Remove Or Use Non-skid Bottom Tape
Garage/Basement/Hobby Areas Locked	☐ Use Keyed Locks
Sharp Implements (Pens, pencils, razors)	☐ Remove, Keep Up Out Of Reach
Swallowable Items (Coins, safety pins)	☐ Remove, Keep Up Out Of Reach
Poisons	☐ Keep Out Of Reach
Poison Labels	☐ Put On Poison Containers
Poisonous House Plants	☐ Identify And Remove
Lead Paint	☐ Have Tested If Painted Before 1980
Radiator (Which can cause burns)	☐ Install Barriers
Stairs	☐ Install Gates Top & Bottom
House Plants (That can be pulled over)	☐ Put On Floor Or Remove
Toy Chest Lids	☐ Safety Stops
Plastic Bags (That can suffocate)	☐ Remove, Keep Up Out Of Reach
Matches, Lighters, Cigarette Butts	☐ Don't Smoke, Keep Up out Of Reach
Guns	☐ Keep Locked Up
Balloons	☐ Keep Up Out Of Reach
Mothballs (Are poisonous)	☐ Don't Use Or Keep Out Of Reach
Insect, Rodent Poison	☐ Keep Out Of Reach
Drugs (Aspirin, cough syrup, prescription)	☐ Keep Locked Up Or Out Of Reach
Kitchen Cabinets (Cleaning supplies)	☐ Use Safety Locks
Bathroom Cabinets (Cleaning supplies)	☐ Use Safety Locks
Hot Water (Scalding, burns)	☐ Turn water heater down below 120°

Section Two:

Building Blocks

Structural Checks

THE BUILDING BLOCKS OF YOUR HOME

The structure of buildings will vary according to region, design, age, and builder or architect's choice. The superstructure rests on the foundation. In some cases, the foundation will be part of the superstructure, such as an historical house with a solid brick foundation and walls. Often the superstructure is covered with sidings and other finishes, which allow for a limited inspection. The components of the structure include: walls, floors, ceilings and roofs.

MATERIALS

Structures are made from various types of materials. These materials will also vary according to region, design, age, and/or choice of the builder, architect, or home owner. These materials may include: wood, steel, concrete, masonry, aluminum and stone.

The application of these materials may also greatly change according to region, design, age, and/or choice of the builder, architect, or home owner. Any of these components, materials or applications have their own unique problems and manufacturer's installation requirements.

WOOD

Moisture and Decay

Wood has been used for centuries for constructing buildings. Some woods, such as fir, locust, cedar, and oak, are much more resistant to decay caused by water and insect damage. Fungus, mildew and dry rot need food, and unfortunately, the entree of the day could be your home. Luckily, would-be diners also need a moisture content of 19% or more and oxygen to survive. Eliminate one of these factors and decay will stop. Usually, the easiest source to eliminate is the moisture, unless you plan on sealing your home airtight, which could prove detrimental to your own health.

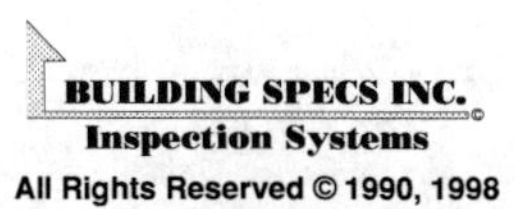

Some forms of dry rot and fungus, once established, can survive purely on the humidity in the air. Levels in excess of 70% will sustain dry rot (despite its arid name) and some fungus. Once the material has absorbed its capacity, it is in "equilibrium."

Moisture cannot be felt by touch until the levels have reached approximately 30%. Since rot can occur at 19%, damage could be occurring to wood when it feels dry to the touch. (A moisture meter would be of assistance in this situation.) What appears to be black mildew could be from the formation, growth and death of the mildew's cycle, therefore the black coloring may indicate an older problem. A moisture meter can indicate if the cycle is active. There are topical treatments which can kill or disrupt the fungus' cycle if you find it is growing throughout your home.

Wood destroying fungus.

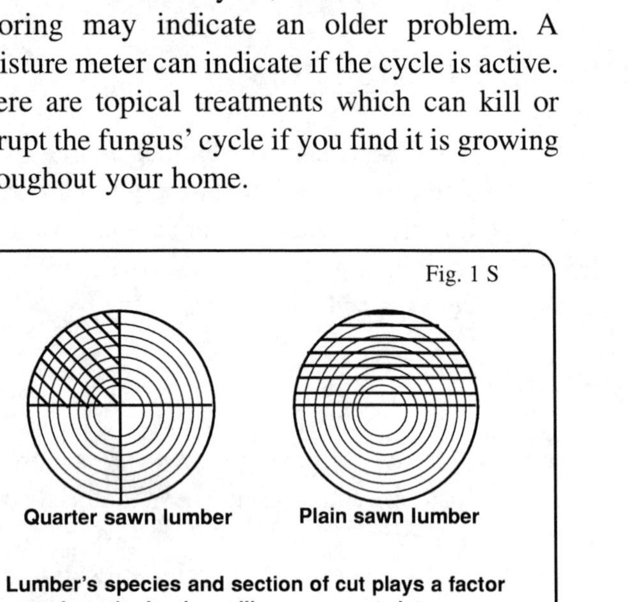

Lumber's species and section of cut plays a factor as to how the lumber will move, cup, twist, warp, check, split, compress, and maintain its shape. Plain sawn lumber has more random cuts. The sections close to the perimeter are much more unstable and prone to physical change.

Wood has two basic classifications, softwood and hardwood. Soft woods are coniferous (evergreens) and used in general construction. These woods include cedar, redwood, fir, pine and spruce, and can range greatly in strength and capacities. Hardwoods are often from deciduous (leafed) trees, and are used for flooring and stairs. These woods include oak, hickory, locust, and maple.

The grain of the wood, how it was cut from the tree, and how it is used, will affect the structural ability of the wood. A board that is of vertical grain will tend to be stronger, and more stable. It will not be as prone to cupping or suffering a raised grain. Plain-sawn boards, on the other hand, are sliced one after another through the log. These boards are more susceptible to cupping and will have more ranging grain patterns. Quarter sawn is a log which has its radius quartered and then cut at a 45-degree angle to the center cut. This cut has more of an even grain pattern, is more stable of a board and has more waste when cut from the tree. (See Figure 1S)

Wood Facts

Wood is subjected to certain stresses when used as a structural member.

- *Tensile strength* is the force pulling at the grain.
- *Compression* is the force being applied to the grain. Example: a floor joist with a waterbed in the center span, the top edge is in compression the bottom edge is in tension.
- *Modules of Elasticity* refer to the board's ability to regain its original form once a load has been applied.
- *Shear* is the force that can cause a board to break across or with the grain.
- *Deflection* is the amount a board will give when under load.

Other Characteristics of Wood

- *Wane*: Sections of a cut of lumber that are too close to the bark. This leaves an imperfection in the wood.
- *Pitch pocket*: Void filled with sap.
- *Knot*: Section where a branch once connected. Possible weak area in board.
- *Shake*: A split occurring between growth rings.
- *Checks:* (run across grain, during drying) Some of these are grading defects, which could affect the integrity of the wood. The size of the board can vary depending on span, spacing of the joist, load applied, and species.
- *Cupping*: curling viewed from the end grain of a board.
- *Twisting or warping*: Viewing from the end of the board, irregularities in various directions.
- *Crown*: Sighting the top edge of a board, the crown is the high point or arched top up.

> ***Definitions...***
>
> **Joist:** Horizontal structural member of the floor.

FRAMES

Wood rot, commonly found at end grain.

Applications associated with wood have ranged over the years. These different frame styles include the following:

Platform Framing (See Figure 2S)

This style of framing is a contemporary method, where a floor system is first built on top of the foundation, and may be built from many components. The floor joist may be wood or steel.

Wood joist can be fabricated from:

- *Dimensional lumber:* When dimensional lumber is used in framing floors, ceilings and roofs, the crowns of the boards should be installed up --- This way, the board's natural camber helps the structural integrity and reduces low spots. When the walls are built, the crowns should all be in the same direction. This helps cut down on severe undulation in the wall.
- *I beams (plywood or OSB webs):* This type of member is very true and stable. There is little or no

deviation from joist to joist. It requires web stiffeners and crush blocks over bearing points and bridging at center spans.

• *Open webbing floor trusses:* pre-engineered wood in typical 2x4 wood triangulation construction.

• *Micro lams:* Pre-engineered, built-up versicle veneer (thick plywood) systems used for beams and excessive spans for floor joists or headers.

> ***Definitions...***
>
> **OSB:** Oriented Strand Board. Pre-engineered lumber; strips of wood bound together with glue under high pressure.

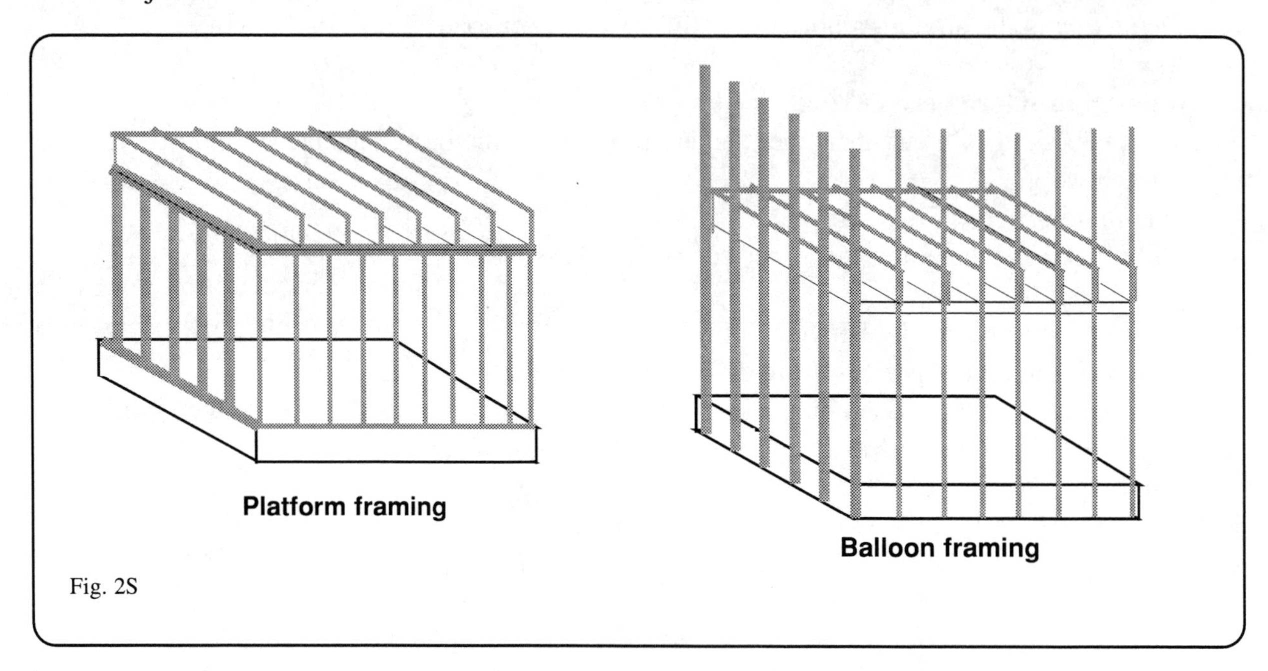

Fig. 2S

• *Glue lams:* Pre-engineered, built-up 2x4's laminated flat on top of each other used for beams and excessive spans for floor joists or headers.

• *Open metal trusses:* pre-engineered metal welded sections utilizing triangulation construction.

• *"C" steel joist:* Thin solid web metal joist. Requires web stiffeners and crush blocks over bearing points and bridging at center spans.

After the floor joists, the subfloor (usually plywood or OSB are used) is installed over the floor joist, and the walls are built over the top of the subfloor (platform). This method is usually considered very stable since the structure consist of all direct bearing loads.

Balloon Framing (See Figure 2S)

This type of framing, used earlier in this century, has some inherent problems. The use of long, straight, wall members is expensive and not practical. With balloon framing the floor joists are attached to the wall, which can be accomplished in a few ways:

1. The floor joists are nailed to the side of the studs. This sometimes is all that holds the floor structure. This method can be hazardous, for the load of the floor is transferred to the nails. If the nail connections to the wall fail, the floor could collapse.

2. One variation of the previous method is to also nail a ledger onto the side of the wall. This

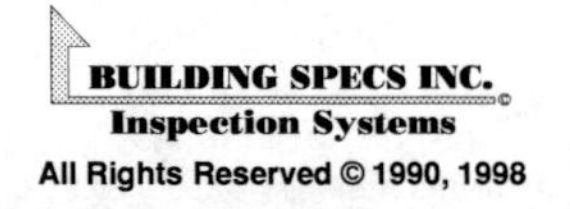

gives the floor joist some direct bearing.

3. This ledger can also be notched into the studs, to increase the strength of the ledger, by transferring the load directly to the studs.

This type of framing needs to be inspected closely. As you walk around the floors, bounce lightly and feel for any excess deflection. Look around the perimeter and see how much the floor moves at the baseboard trim. Floor joist at foundation level usually rest directly on the foundation wall, and the studs come down alongside the joist. This straight chase can pose a problem in the event of a fire. The flames can quickly run up the drafty chase and consume the building.

Stick Built

This refers to a house being built on site, from scratch.

Panelized

Panelized homes have sections of the walls prefabricated at a factory and erected on site. Some walls may be set in place by hand, others may need a crane. Usually, the walls have only the exterior sheathing on them, though some manufacturers will have wall sections with drywall, insulation, wiring, siding, windows, and other components already installed.

Modular

Modular homes are prefabricated large sections craned in and set on a foundation. The walls usually have all or most of the components in them. Once the house is set in place, some joining of the sections is required, which might include drywall, taping, siding, and roof capping. Houses more elaborate in design might require some framing and additional construction. The plumbing, heat, and electrical systems have to be connected as well. Some simpler ranchers might come in two sections and need to be joined.

Good quality modular homes do not sit outside with their framing exposed to the elements. Usually the buildings are completed within a set allotted time, and are not as prone to delays as stick built.

 ***Problems to look for: MODULAR HOMES***

- In some states these homes are not regulated by the county codes but by state building codes.
- Hinged (swing) trusses (pre-fabricated roof system) are susceptible to weak connections at the hinge. The amount of triangulation is limited, and these areas may need additional bracing.
- Alignment problems at installation, which are repaired in the field.
- Inferior construction.

Log Cabin

Log cabins can date back quite a few years. Older cabins can even have mud and grass chinking between the logs, but more than likely your home is a little more modern.

Later model cabins use a cement based chinking, and now modern log homes utilize grooved logs and a foam gasket between the logs. The logs are also treated or built from more decay resistant materials.

 ***Problems to look for: LOG CABINS***

- Inadequate foundation or ground contact with lumber.
- Rot at ground contact, around cuts in the walls (windows), ends of logs.
- Inadequate insulation.
- Possible airborne pollutants from treatments to logs.

Post and Beam

Post and beam homes are one of the strongest, longest lasting wood structures built. The superstructure is made from large timbers which are notched and fitted together and are usually able to support themselves without the aid of fasteners. The structure has triangulation and supports built into the system, and fasteners and pegs are used to keep everything tight. Once the superstructure is erected, the walls are filled with smaller sections to fill out for siding and drywall.

Post and beam construction is enjoying a renaissance. The building technology is more advanced, and the blueprints lay out the specification to 1/32".

Older post and beam homes (historical) were built from hand-hewn lumber, where timber is flattened by an adz (shaving axe). In some buildings you can still see round sections of the tree and even bark. In these older homes, uneven floors, roofs and walls are typical. However charming this may be, if visible, check all of the joinery for proper connections.

The biggest problem with these buildings is poor maintenance and persistent water problems, which can lead to the structure's premature demise. These homes can also be difficult to blow insulation into, due to the cross bracing in the structure.

 ***Rot Trouble Spots with Wood***

- *At all doorway entries.* Tap plywood and listen for the hollow thud of soft or delaminating sheathing. Insert sharp probe at intersection of jamb and floor to check for soft rot spots. Moisture and insects typically infiltrate around the exterior casing or at the exterior toe kick.
- *At door entries* on decks or concrete stoops where door threshold is within inches of the surface. Water and snow can back up and seep under the threshold.
- *Lack of drip caps* over wood trimmed windows or doors lets water infiltrate behind.
- *Vertical trim that runs to the ground* or concrete where the capillaries in the wood can absorb the water. This is referred to as wicking and is very common at garage doors and entry doors.
- *Any siding that is close to or touching the ground.* This can cause damage to wood siding as well as to the sheathing and framing in the wall.
- *Columns or post resting on concrete without a pad* can soak water into its capillaries.
- *Roof sheathing* due to ice damming and leaves clogging, and also where there is no drip edge at the gutters.
- *Tongue and groove porch flooring at the end grain.*
- *Closed in porches with the decking still exposed* to the outside.
- *Water stains or rotting at the sheathing from the end grain* in the attic.
- *Sheathing* at plumbing vents, around chimneys, and around roof penetrations in the attic.

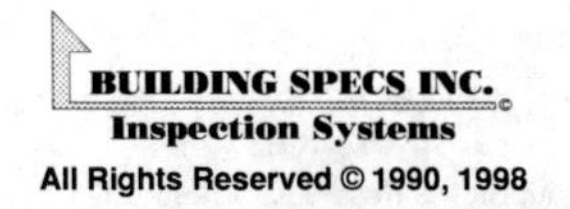

- *Wood exposed* to the south side or large amounts of direct sun.
- *Wood exposed* to continuous dampness.
- *Damp or wet crawl spaces* can promote fungus, mildew and lead to rot.
- *Any wood trim* especially at the end grains.
- *Pine and spruce which are not back-primed* (sealing the back edge of a piece of trim) or sealed at the endgrain.
- *Any plywood* used outside for paneling or trim that is not marine grade (premium cut of lumber and moisture resistant glues).
- *Sill plates* that are untreated.
- *Floor sheathing* at toilet drain and other plumbing floor penetrations.
- *Wooden window sills.*

As solid wood becomes a disappearing resource, products such as processed wood (OSB), plastic composites (newer decking materials), and metal is going to become more commonplace. With the introduction of these materials comes different installation techniques. New installation criteria will be developed as problems present themselves, and there will come an evolution of material, knowledge and skill.

METAL

Metal framed buildings have been constructed in southern regions for years. The popularity of metal as a framing material in these climates is due to the high risk of termite and rot damage. Metal studs were more prevalent in commercial construction, but are now popular in residential buildings.

As the use of metal studs spread to the northern regions, some problems appeared. Many problems are caused by carpenters and builders who do not possess the proper education or training regarding the proper installation methods for metal frames.

 ***Potential Problems: METAL***

- Proper fastening to areas such as roof trusses is necessary using techniques such as metal tie straps or spot welds. If the trusses are merely screwed to the top plate, high winds could lift the roof system from the house.
- Exterior and load bearing walls should have a heavier gauge steel channel than non-load bearing walls. Bridging in the wall cavities can help to prevent stud movement.
- Floor systems which use channel steel require bridging to prevent racking and crushing to the web. Racking of the floor joist should be taken very seriously. A building could shift and collapse within the floor system.

Metal floor joists. Look for bridging and web stiffeners.

Bridging should be used in the center of the spans and below load bearing points, such as where a load bearing wall rests on a main girder. Another area to reinforce would be a parallel wall bay. This could cause sagging between the joist unless properly supported. This can be supported by installing bridging 24" oc., or installing two parallel joists offset from the wall, one on each side.

Definitions...

Bridging: Perpendicular framing member installed between joist and rafters at center span to help distribute load and prevent racking.

Thermal Bridging

Another problem with metal frames is thermal conductivity or thermal bridging in the northern regions. Outside air temperature is transferred through the metal stud to the inside and warmer air, causing condensation inside the wall, evident by streaking showing through the drywall. The streaks are lines reflecting the location of the metal framing in the wall, and may be vertical or horizontal. Frost and mold can form inside the wall cavity as well.

Metal conducts heat and cold 300 times faster than wood. This drops the R-value of the wall significantly. There is research being conducted on proper remediation and installation techniques to deal with this issue.

This thermal bridging problem can be aggravated by using hardboard wall sheathing such as oriented strand board. One recommended procedure is to install extruded polystyrene foam over the studs on the cold side to reduce the chance of a condensation problem. This layer of foam is referred to as a thermal break. However, even with foam covering the studs, the fasteners can still conduct the cold to the stud and encourage condensation.

Another method of prevention currently under scrutiny is the use of horizontal "hats." This is a channel which runs across the studs which breaks the thermal bridge. Some companies have developed studs with open webbing, as opposed to solid, which are supposed to significantly reduce thermal bridging. There is other ongoing research, including the use of wooden washers.

Any areas susceptible to air infiltration within metal framework should be caulked or spray foamed. If these conditions are not rectified, the condensation could lead to rusting and to the possible failure of the metal studs and the walls. Further evaluation and monitoring is recommended.

Eventually, a system will be devised so that metal studs and applications will have guidelines to prevent such problems. A professional should be consulted if you suspect a problem.

Skeletonized Beam Frame

Another method of steel construction is the use of a skeletonized beam frame. Once the superstructure is erected, the sections between the beams are filled with lighter gauge metal studs.

Steel trusses are used for floor joists and roof rafters, as they have an open web using triangulation for support. This allows for an easy access for plumbing and electrical chases. This type of structural member is more common in commercial construction, but is occasionally used in residential. Long free spans with no center support can be achieved with this type of truss.

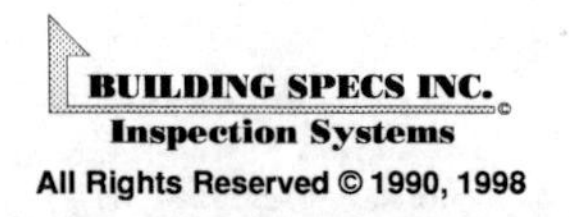

ROOF STRUCTURES

The roof structure is the frame which supports the weight of the sheathing, roofing material and other loads. These loads may include a deck, heat system, solar panels, and other items. If the building is in a high wind potential area, there are hurricane straps available, which tie the roof structure to the wall. This helps to prevent wind from lifting the roof off the building.

The structure can be broken down into a few styles.

Trusses

Trusses are pre-manufactured or site built, usually self-supporting with triangulation built in, and may be constructed from wood, steel, or concrete. There are a few styles of trusses, however, most consist of a bottom cord, top cord, and web system. A truss can be flat or have a sloping pitch, and can be designed to stage down for use with hip and valley roofs. At the points of intersection, modern trusses have a pressed-on steel plate. Older trusses may have been site built using metal plates or plywood secured with nails. Some special circumstances may require that same application today, however they must be approved by a structural engineer.

Wood trusses are sometimes installed out of alignment, which throws off the ridge and alignment of the slopes. Look for any broken or damaged cords or damaged connecting plates. Trusses allow for less storage and available space in the attic due to the web.

Rafter

The rafter refers to the member that makes up the actual slope or body of the roof. This needs to be sized according to pitch, span, spacing, and other loads to be applied. For example, a low pitched roof could sag under snow load, whereas a steeply pitched roof may not sag under the same conditions, since it transfers the weight differently to the structure.

Rafters are used when the roof is stick built on site and may be built with a few variations. First, the roof structure needs to be supported in some way, and this can include triangulation at the ceiling rafters or collar ties. Another method is to create a self-supporting ridge with a wooden or steel beam. Region and age can be a factor in what method the builder chooses.

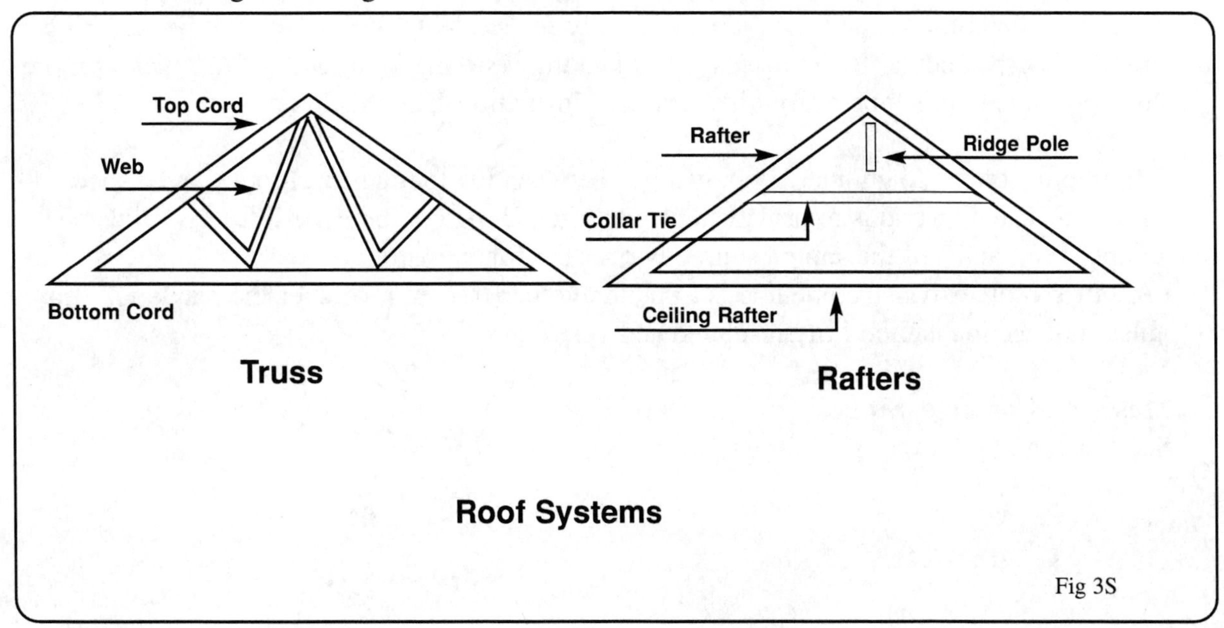

Fig 3S

> ***Definitions...***
>
> **Collar Ties:** Horizontal member used to create triangulation in rafter systems.

Some rafter roofs will have a ridge pole at the peak of the roof made from a 1x6 up to a 2x12 board or even a wooden or steel beam, helping to tie the structure together. Some older homes do not have a ridge pole and merely butt the tops of the rafters together. When spaced slats are used with this type of ridge, diagonal support may be necessary to the rafters. When inspecting the cuts at the top where they intersect adjacent rafters or the ridge pole, look for excessive gaps which may indicate settling or movement. For example, several rafters with a 3/8" gap at the bottom of the cut adjacent to the ridge pole and tight at the top may indicate inadequate triangulations at the walls and possible settlement to the roof structure.

The collar ties or beam creates triangulation, which prevent spreading of the walls when a downward load is applied to the roof. Collar ties also help with side forces such as wind against the roof, by transferring the load.

 ***Problems to look for: RAFTERS***

- Sagging roof, bouncy structure (inadequate sized lumber).
- Bowing exterior walls or a sagging ridge could indicate failing connections at roof rafter/ceiling rafter, lacking collar ties, foundation settling, failing floor system, amateur construction.
- Hump (poor framing alignment).
- Cracked rafter.

Some of these conditions can be stabilized with knee walls, bridging, or sistering another section of wood to the a weak or inadequate rafter. However, some of these repairs require extensive evaluation and work, and a professional should be consulted.

Wood Plank and Beam

This style is often used in contemporary buildings with exposed wood ceilings. Appropriate beams are installed further apart than rafters, usually several feet. Tongue and groove planking is then installed perpendicular to the beams. The planking is normally at least 1-1/2" thick or more.

One downside to this type of roof is method of insulation. They include:

- None
- Rigid polystyrene foam boards sandwiched between the tongue and groove and a sheet of plywood. The plywood is to provide a place for the shingle to be nailed. Cheap builders skip the plywood and nail the shingles directly over the foam board.
- Another, more expensive, method is to build another framed roof over the planking. This false roof section is then fully insulated and ventilated.

Masonry, Concrete, Stone

See Foundation.

STRUCTURE CHECKLIST

1] **WOOD** **Framed Walls** ☐N/V ☐2x4 ☐2x6 ☐2x8 ☐Engineered Lumber ☐Other ____________

a] **Style** ☐Platform ☐Balloon Framed ☐Stress-Skin Panels ☐Log ☐Post & Beam ☐Other ____________

☐Ok ☐**Problem**[1] ☐Minor ☐**Severe / Amateur, Sloppy Workmanship / Signs of Settlement[1] / Water, Insect Damage / Rot**

b] **Floors** ☐N/V ☐2x8 ☐2x10 ☐2x12 ☐I Joist ☐Truss ☐Beams ☐Other ____________

☐Ok ☐**Problem**[1]; **Sagging / Bouncy / Hump / Visibly Out of Level[1] / Signs of Settlement[1] / Water, Insect Damage / Rot**

Floor Level ☐**#1** ☐**#2** ☐**#3** ☐**Stair Landing(s)** ☐Minor ☐**Severe**[1] ☐**Monitor and/or Repair as Needed**

2] **MASONRY** ☐Ok ☐**Block** ☐**Poured Concrete** ☐**Stone** ☐**Brick** ☐**Terra Cotta** ☐**Coral**

a] **Settlement Cracks**[1] ☐N/V ☐Minor ☐**Severe** ☐**Extensive** ☐**Monitor and/or Repair as Needed**[1]

Stair-Stepping / Vertical / Horizontal / Loose Brick, Block / Deteriorating / Failing / Bulging / Frostline Crack / **Amateur / Sloppy Work**

b] **Water Damage**[1] ☐N/V ☐Minor ☐**Severe** ☐**Efflorescence** ☐**Deterioration / Crumbling** ☐**Previous Repairs**

3] **METAL** ☐**Studs** ☐**Joists** ☐**Trusses** ☐**Rafters** ☐**Structural Beams, Frame** ☐Other ____________

☐Ok ☐**Problem** ☐N/V ☐Minor ☐**Severe;** Mildew Streaks / Rusting / Requires Thermal Bridging / Improper Installation

•**ADDITIONAL COMMENTS** *[1]Any problems should be monitored and repaired as needed by a licensed contractor or qualified individual. Severe problems such as cracking, amateur work, water, and/or insect damage could lead to a structural failure.*

☐**Further Structural Evaluation and/or Repairs Required**[1] ☐**Not Visible or Accessible For Inspection**

Foundations

"A house is as strong as the foundation it is on" isn't always just a metaphor. The foundation is a significant factor in whether or not a house is susceptible to settling, or if water will penetrate and leak into the basement. The quality and proper use of materials is as important as the workmanship. Unfortunately, the foundation may consist of several components, and many of these components may be hidden behind finished walls or below grade, impossible to inspect.

FOOTING

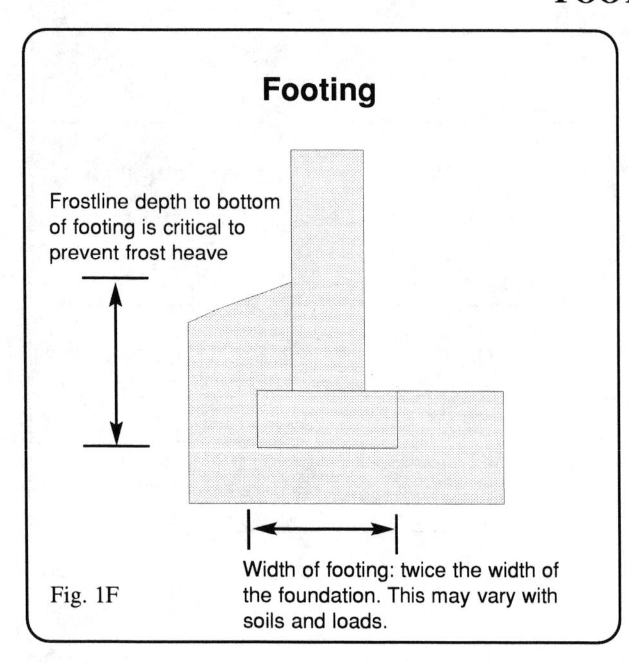

Fig. 1F

The footing is the most common base used for foundations. Most modern solid foundation walls consist of a solid perimeter pour. The footing is always dug to an appropriate depth and width, but otherwise varies due to several factors. For example: Homes built on floodplains may require piers (independent vertical supports constructed from wood, concrete or masonry), and historic buildings were commonly built on stacked stone foundations, rather than solid perimeter pours. Size and depth of footings may vary with climate, region and soil composition.

Soils vary in their ability to support the weight of the structure. Clay, sand, low lying wetlands, coral, stone, bedrock, etc., all have a different weight bearing tolerance, so wider footings may be used to distribute weight over a greater area in the soil. This will help prevent a point load and reduce the chance of settling. A building is more stable and less likely to settle on a coral or stone base, as opposed to a low lying wetland. In extreme cases of soft soil, the footing may have to rest on wooden pilings, which may be driven into the ground with a piledriver. A soil engineer counts the number of strikes and notes

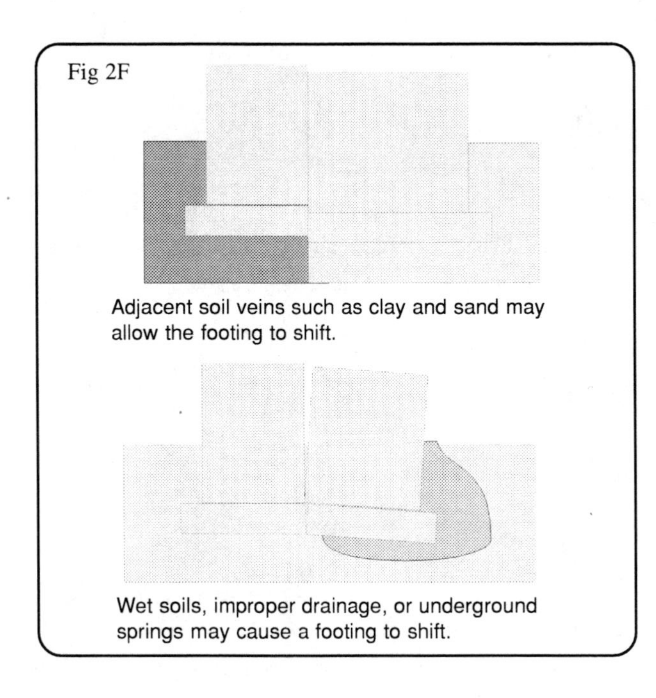

Adjacent soil veins such as clay and sand may allow the footing to shift.

Wet soils, improper drainage, or underground springs may cause a footing to shift.

at what rate the piles went down per strike. Once pilings are in place, a continuous footing is poured on top.

Another factor affecting the size of the footing is the depth of the frostline. This is the depth that the ground will freeze. If frozen ground is allowed to get under a footing, or push against a wedge shaped footing, the result can be the foundation lifting, a phenomenon known as frost heave. The frost line depth varies in each state and even in each county.

Most footings are a continuous pour of concrete. Metal rods known as rebar are set in place near the bottom third, running with the footing to offer support. Footings may be stepped up into a hill or sectioned off to accommodate piers or point loads. Interior columns in a basement or crawlspace should have independent footings. They should not rest on a slab alone, unless design was factored in.

If a footing is on two different veins of soil such as clay and sand, the footing could shift and allow other sections of the structure to settle. Cracking in foundation walls is sometimes attributed to footings undermined by erosion or a shear break. Rebar can help prevent this from occurring.

Of course, there is no way to inspect the footing after the house is built (unless it is excavated --- but you probably will not opt to start digging up a home.) Underpinning methods are available to help stabilize settlement that may be occurring. This can be done by installing an additional footing below the existing footing at key points. Metal augers and other specially designed devices to underpin footings are available.

Affects of improper drainage undermining foundation.

Some older buildings may rest solely on a stone laid footing. If there has not been any settlement, and it is an older house, chances are good future settlement may be minimal.

FOUNDATION

There are several types of materials used for a foundation. They include:

- Block
- Brick
- Poured Concrete
- Precast Concrete
- Stone
- Coral
- Wood/less common materials.

As you walk along the foundation look for cracks, which could indicate an existing or older problem. Types of cracks to look for:

- Horizontal (frostline or hydrostatic pressure crack)
- Vertical
- Lateral
- Stair-stepping

A crack may happen in the first few months after a building is erected, or may develop over a period of time. Foundation or masonry cracking or deterioration may be caused by:

- Shifting in the soil.
- Normal settling.
- Undermining of the footing due to poor drainage.
- Inadequate footing.
- Inferior materials.
- Hydrostatic pressure, excessive moisture.
- Ground freezing.
- Expansive clay soils, (shrink, swell).
- Freeze thaw cycles due to expansion and contraction.
- Thermal expansion, where the surface has heated up faster than another area.
- Varying expansion coefficients.
- Remodeling, i.e. adding a second floor.
- Pollutants, i.e. acid rain.

Steps may be taken to diagnose the degree of a problem, prevent further settling or damage, and to implement repairs.

Problems can occur from remodeling if proper care is not taken. For example: putting a second story on a house which has inadequate footing could cause shifting; or building a garage adjacent to a foundation wall could put lateral pressure against a foundation wall, which could lead to a foundation wall cracking inward.

Any crack should be monitored, and there are a few methods you can use. The first method employs a device with a grid and a dot in the middle. This device is epoxied to the crack, and the movement of the dot is noted over a period of time, usually several months. Another method is to draw a horizontal line across the crack with a thin waterproof marker and then take a caliper and measure the width. Look for any by-passing of the lines (shearing) or any lateral movement (indicated by bulging of the wall). An alternative is to repair the crack and monitor it for activity. If the crack reappears, it may indicate movement.

Depending on the severity of the crack, the degree of repair will vary. Minor cracks may require only point up. Point up of cracks should be done with an epoxy, a polyurethane, or other similar sealant. Standard mortar will usually work loose and fall out. Silicone and other standard caulks will not bond to the masonry.

Larger, more severe cracks may be filled with a similar sealant or a hydraulic cement. Preventing moisture seepage through the crack may be a secondary problem.

Remember that there is always the chance that the foundation wall may have been structurally weakened by the movement which caused the crack. In this case, further structural evaluation and repairs may be necessary.

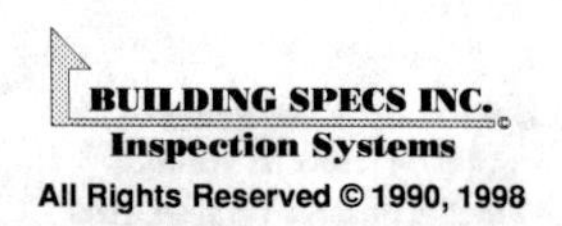

Concrete Block

Concrete block is susceptible to stair-step cracking, which is evident by both horizontal and vertical cracks which appear in the mortar joint. This is usually attributed to normal shrinkage and/or settling or thermal expansion. If the cracking continues to enlarge, the cause should be determined.

Horizontal cracking in the mortar joints may be due to wire wall ties laid in the mortar bed causing uneven expansion and contraction. Horizontal cracking that is sagging or perpetually growing may be due to more serious settling. Another form of horizontal cracking can appear in conjunction with a bulge in the wall. This may be due to hydrostatic pressure from the soil, and ground water pushing against the wall. This may also be caused by a freezing of the soil which causes a frostline crack. Since the building and footing hold the wall at fixed points, the center may weaken due to the external pressure. To detect this, hold a straight edge, preferably a 4-6' level against the wall. See if the straight edge rocks over a high spot at the crack. With the level in a plumb position measure the distance off the wall to the level, and mark the exact location on the wall for future reference. This may need to be done in a few areas over several months to evaluate whether or not there is any significant movement. A severe case could lead to a wall caving in; it should not be taken lightly.

Another crack that needs close attention is a vertical shear. This may indicate that something has definitely shifted, such as a footing, and should be monitored and repaired as needed. Underpinning the footing may become necessary.

Concrete block itself can deteriorate. You can test the block by scraping the exposed surface with a screwdriver or similar probe. If the surface is sandy or crumbles, the block may be failing. If the block has failed to the point of completely cracking, immediate repairs may be necessary.

Signs of previous termite treatment: drilled and plugged holes in concrete.

Termite Treatment

Signs of previous termite treatment may be observed at concrete slab, block, or poured foundation, evident by drilled and plugged holes. In southern states a manifold may be incorporated into concrete block, and the pesticide sprayed directly and routinely. In some southern states certain termites can actually destroy masonry, concrete, drywall, lead and copper. See additional information under the "Basement and Crawlspaces" chapter under "Insect Damage."

Parge Coat

A block wall may have a texture coat of cement spread over it, from the footing to the siding, known as the parge coat. A parge coat can be used for aesthetic value as well as for a moisture barrier. A coat of tar should be applied to the parge coat from the grade to the footing to further prevent moisture from passing through the wall. In addition to the traditional tar coat, new water proofing barriers are available. Check with your local masonry suppliers or waterproofing companies.

Look for cracking and loose sections in the parge coat. Minor spider cracks are more of a cosmetic problem unless they spread and widen. Cracks in the foundation may extend through the parge coat. These can be easily repaired, but if the cracks persist or worsen, the cause should be diagnosed and stabilized.

Loose sections can be located by tapping the stucco with your fingers and listening for hollow sounds. These may indicate moisture infiltration. Loose sections or pockets may also be the result of improper application. Water creeping behind the parge coat can compound the problem.

Concrete

A poured concrete foundation is normally very stable and solid. Since it is less porous, it is less likely to develop leaks than some of the other foundations. When a poured foundation develops a crack, it may be attributed to a bad mix, to the fact that it was poured in freezing conditions, or to a severe shift in the ground and footing. Check to see if the concrete is flaking or crumbling. There are tests which can determine the amount of failure in the concrete.

Brick

Brick possesses similar traits to block, with a few additions. Brick may be used as a veneer over another surface like block or wood. Brick may also be used as a foundation. The majority of brick foundations will be found beneath older homes, and usually laid three to five courses across. There is normally no way to inspect the condition of the brick, or mortar, on the interior sections of the wall. All you can do is inspect the visible outside conditions of the mortar and brick itself.

Older bricks may be failing, evident by crumbling and deterioration, due to water damage. In severe cases the brick turns to powder and completely disintegrates. Check to see if the mortar has deteriorated. When the surface of the brick cracks or peals off in a layer, this is referred to as spalling. Spalling may result from the pressure of salts and water seeping into the brick. As the brick freezes and thaws, the brick face may crack. Spalling may also occur due to incompatible mortar. Portland cement is very hard, and when used with softer brick, can cause the edges to crack. As the brick expands and contracts, the hard mortar does not give, and the brick face may crack. Lime is a softer mortar and is recommended for use with older or softer bricks, since lime mortar gives with the expanding and contracting bricks. Proper mortar must also be used when repointing. There are methods available to diagnose the mortar content.

A brick wall which is separating from the structure may require anchors back into the structure. Any bulging or abnormalities should be monitored and evaluated by a professional.

Stone

Stone foundations are more common in older homes. Frequently, the stone is merely dry stacked and has no mortar in the joints. This may be acceptable if there is only a crawlspace; the air spaces can actually promote proper ventilation. However, if there is a water problem, the gaps may need to be filled with a waterproof mortar, or even a hydraulic cement. If the stone goes down to a basement, a serious water problem could occur. In some cases, a concrete reinforced wall is poured on the inside to encase part of the stone, creating a solid wall, and preventing water intrusion between the stone.

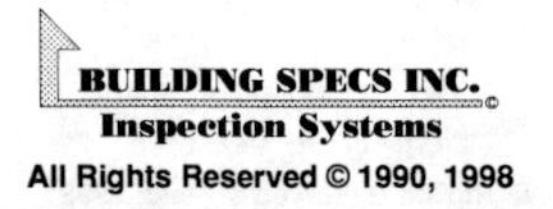

Mortar

Mortar is a mixture of lime, cement and sand used to bond masonry units together. The mortar allows the masonry to breathe. Various mixtures are used; the type used depends on the type of masonry.

During your inspection, inspect the mortar itself. Run your finger through the joint; it should be solid and smooth. If sandy powder falls from the mortar, closer scrutiny is in order. Crumbling, sandy or loose mortar may be due to a poor mix, freezing when curing, age, or hydrostatic pressure water damage. If the the mortar is failing, it may need to be dug out and pointed up.

Mortar is made up of three basic components:

- *Water*
- *Aggregate*, usually sand. Areas where beach sand was used may contain a high salt content and lead to the mortar failing. Some areas even utilize crushed shells, animal hair, and clay in the mortar for an aggregate.
- *Binder, lime and cement.* Lime is preferred when pointing historic buildings, since it is a softer mortar. Small amounts of portland cement may be added depending on the strength of the masonry unit. Portland cement has been used since the end of the 1800's; it is very hard and does not absorb water easily.

Care must be taken not to use a mortar or sealant which is harder than the brick or masonry unit. If the mortar is too hard the masonry unit may crack or chip at the edges as a result of expansion and contraction. This is a common problem with point up of historic buildings. The use of portland cement is not recommended for historic building due to its hardness.

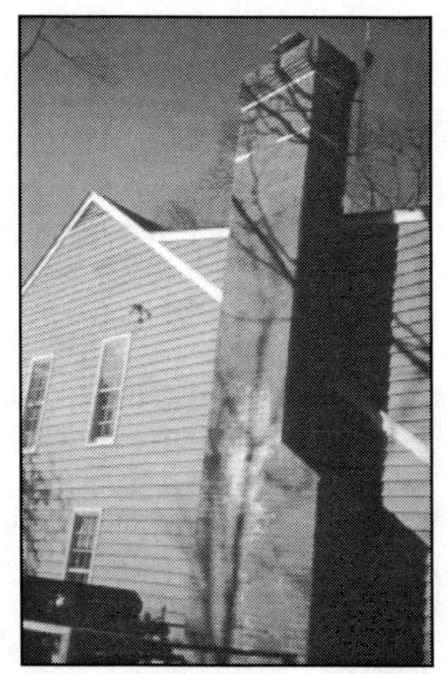

Efflorescence, as seen on a chimney.

Efflorescence

Efflorescence is a salt deposit on the face of the masonry unit. It may indicate moisture behind the wall or rising through the ground. A new home may leave a salt residue from the moisture due to the initial construction. This is referred to as "bloom," and will disappear with time. To see if the efflorescence is active brush it off and monitor it for return. If it returns there is a high probability moisture is entering the wall. This should be stopped to prevent damage to the masonry and mortar; the salt can heat up and lead to damage to the masonry unit.

Wood

Wooden foundations are usually constructed from pressure treated wood and are normally constructed from 2x4 or 2x6 lumber. The wall is then covered with pressure treated plywood. The foundation wall can sit on a concrete footing or crushed gravel base.

Wooden foundations are susceptible to rot and deterioration if the water barrier fails. There is no way to know the condition of these walls without excavating the foundation. If there are signs

Deteriorating brick pier due to water damage.

of moisture penetrating the wall inside, further evaluation is in order.

General Foundation Tips & Parts

When inspecting a basement, see if it was dug out after the house was built. This may be evident by an offset in the wall where the footing is protruding inside. If the ground has been dug close to the footing perimeter and no retaining or lateral support is present, an engineer or appropriate individual should be consulted to evaluate the situation. This can become a serious problem due to pressure on the soil. If the dirt fails and begins to fall away from the foundation, the footing could shift out from under the house.

While you are walking around the house, take note of any ventilation. Ideally, you want to see cross ventilation. If the vents are adjustable, check to see if they are operational.

Lintels

There are two common types of lintels, steel and concrete, both of which are used to span the opening across windows and doors. Steel lintels are subject to rust, and should be painted with a rust inhibiting paint. With concrete lentils, look for cracks to the lintel itself. Minor cracking at the wall connection may be normal, due to expansion and contraction.

Piers

Some houses are built on piers, columns that are spread accordingly under the house. Piers are typically constructed from wood, block or concrete.

Though piers can range in appearance from something resembling a telephone pole to an 8'x8" timber, all wooden piers should be treated with a ground contact rot inhibitor. In any case, you will want to look for possible rot, especially at ground level. Look for any diagonal bracing to prevent swaying to the building. Be sure the joist system is securely bolted to the piers and inspect the piers closely for any severe warping or splitting.

Block piers should be poured solid and have a positive connection to the footing with rebar. These piers also require adequate footing, which may vary, due to soil conditions. Soil compression may be noticeable if you see the concrete footing sinking below the surrounding soil. Piers that are not poured solid are acceptable, but may be prone to some problems, including separation from the footing. The best method is to have the pier poured solid and attached to the footing with rebar. Check for plumbness, settling, bulging or leaning. Also check to see if there is a positive connection to the floor system.

Concrete piers can be tubular shaped if originally poured into a sona tube (prefabricated cardboard form for cement). They can also be be poured with a cantilevered top for added support. Examine the piers for cracks, flaking concrete, signs of deterioration, leaning, bulging, or settling.

When inspecting piers, it is important to verify that an adequate number of piers are present. Look for signs of sagging at the girder beams between the piers. If evidence of a problem exists, you may need to consult with an engineer to determine proper spans and loads.

Cleaning Masonry

Below are some products used to clean masonry. When using any of these products, always experiment on a small area for reaction to the masonry unit. Stains, discoloration or damage may occur, so it is always a good idea to consult a professional before using any cleaners.

- *Water Washing:* hand washing, power washing, spraying with a hose.
- *For Graffiti, Spray Paint:* try acetone and other solvents.
- *For Rust:* oxalic acid, sodium citrate.
- *For Climbing Plants, i.e. ivy:* herbicides, detergents, peroxide, bleach.
- *For Excessive Dirt:* muriatic acid, hydrofluoric acid, hydrochloric acid (not on limestone, marble, sandstone).

FOUNDATION CHECKLIST

1] **TYPE OF MATERIAL(S)** ☐Block ☐Poured Concrete ☐Slab on Grade ☐Brick ☐Stone ☐Wood ☐Terra Cotta

a] **Settlement Cracks**[1] ☐N/V ☐Minor ☐**Severe** ☐**Extensive** ☐***Repair and/or Monitor as Needed***[1]

Stair-Stepping / Vertical / Horizontal / Loose Brick, Block / Deteriorating / Failing / Bulging / Frostline Crack / *Amateur, Sloppy Workmanship*

b] **Water Damage**[1] ☐N/V ☐Minor ☐**Severe** ☐**Efflorescence** ☐**Deterioration / Crumbling / Previous Repairs**

2] **PIERS** ☐**Yes** ☐N/A ☐Block ☐Brick ☐Concrete ☐Stone ☐Wood ☐Steel ☐Other________

☐**Problem** ☐N/V ☐Minor ☐**Severe**[1] ☐**Leaning / Failing / Rusted / Rotted / Inadequate / Appears to be Sinking**

3] **MORTAR JOINTS** ☐Yes ☐**Problem**[1] ☐Minor ☐**Severe** ☐**Mortar Joints Require Point Up**[1] *Use proper mortar, epoxy or polyurethane caulk*

Cracking / Sandy & Crumbling / Loose / Efflorescence / Failing / None / Amateur, Sloppy Work / Not Visible / Soft Brick Hard Mortar Repair

4] **SLAB ON GRADE** ☐Yes ☐N/A **Problem;** Possible Water Infiltration / Damaged / Settling / Undermined / Hollow Soundings / No Footing

5] **LINTEL(S)** ☐N/A ☐Steel / Concrete / Other ☐**Expansion Cracks Noted** *(Point up cracks and monitor, may be only cosmetic)*

☐**Problem;** Rusted / Needs Painting / Cracked / Damaged / Failing / Recommended / Missing ☐***Structural Repairs Required***

6] **SIGHTING WALL** ☐Ok Irregular / Bulging / Leaning ☐Minor ☐**Severe** ☐***Further Structural Evaluation and/or Repairs Required***[1]

7] **PARGE COAT** ☐Yes ☐N/A ☐**Problem;** Surface Cracks / Separating From Wall / Crumbling / Recommended / Loose Pocket(s)

8] **WATER BARRIER** ☐Yes ☐**N/V** ☐N/A ☐No ☐**Recommended** ☐***Foundation Water Proofing Recommended***

•**ADDITIONAL COMMENTS** [1]*Any cracks or water damage should be repaired and monitored as needed by a licensed or qualified individual. Severe problems such as cracking, amateur work or water damage could lead to a structural failure. Any cracks need to be monitored to determine if active or from previous movement.*

____

____

____

____

____

____

____

____

____

____

____

____

____

____

____

____

____

____

____

____

____

☐**Limited / No Access** ☐**Previous Repairs** *(Monitor for further activity)* ☐**Further Structural Evaluation and/or Repairs**

Basements & Crawlspaces

An unfinished basement or a crawlspace can show you a great deal about the condition of your house, such as its quality of construction, visible settlement, and often other unforeseen problems. An unfinished basement will allow the opportunity of a more thorough evaluation, unless the walls are covered with insulation. In this case, try to pull apart any seams in the insulation to see the condition of the foundation wall for cracks or moisture. There are several things you will want to look for once inside. What you inspect for will vary from house to house, and the location and age of the house will also play a factor in how the crawlspace is built.

CRAWLSPACES

Access

Crawlspaces should have an access panel which may be found in a few different locations. Look for them around the perimeter of the foundation, inside of a closet at the floor, or at a basement or other interior wall. Inspect the panel for damage and proper working condition. The access should not be located where it may susceptible to water entry. If there is no access be sure to make a note of it.

> *WARNING!*
> **Wear a respirator when under homes or in crawlspaces, there will be dust, molds, fungus, and insulation, any of which could be harmful.**

Before entering a crawlspace, you should assess the access opening for size and safety. Also, determine the height from the ground to the bottom of the floor joist inside the crawlspace. You don't want get stuck in a tight area. Adverse conditions may warrant a reinspection once the conditions are improved, or prohibit access at any point in time due to restrictions. You should enter a crawlspace with coveralls, flashlight, probe, (moisture meter if available), and a respirator. Look for the presence of animals such as rodents and snakes, which might not be happy to see you. Spiders, scorpions and other biting and stinging insects often hang from the floor above, so be wary. In general, crawlspaces are not the most glamorous places to be.

Ground

Crawlspaces may have dirt, gravel, mud or concrete floors. In most cases a vapor barrier should be on top of a dirt or gravel base. Some crawlspaces may contain sandy soil which will

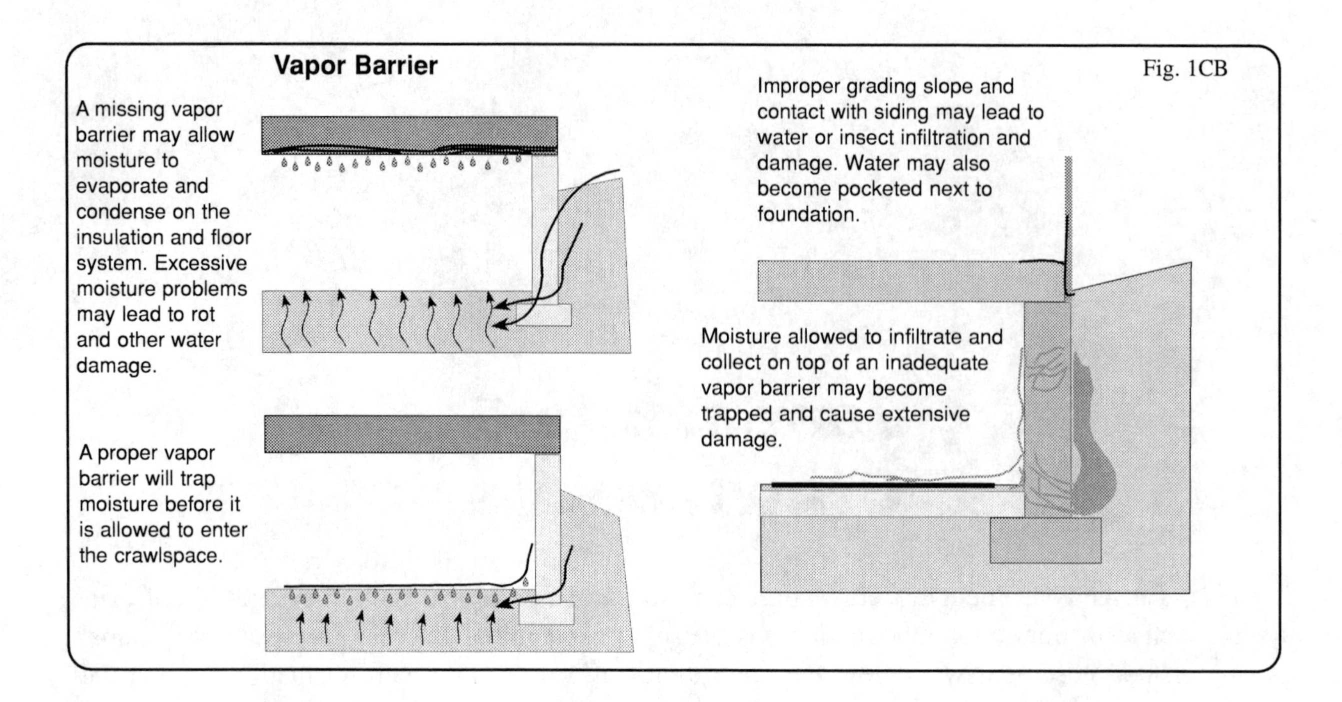

drain well and not require a vapor barrier. Some builders put the vapor barrier under the gravel; this can allow for holes in the plastic and defeat the purpose of the barrier. The vapor barrier should overlap at the seams by several inches or be sealed and draped up onto the wall above the weep holes and any seepage. If moisture ever becomes trapped on top of the vapor barrier it will not be able to drain. A crawlspace with a slab would have the vapor barrier below the slab. Be sure the weep holes are not visible on top of the slab.
(See Fig. 1CB).

Overall water problems should be inspected at various areas throughout the crawlspace. Check the ground for standing water or dampness. Look for any signs of erosion which could indicate a past or recurring problem. See if any erosion is undermining any footings or piers. Trenching for water control is not recommended: over a period of time erosion may result.

Look for any standing water or severely damp ground which could result in an electrical shock. If there is water, and you still want to gain entry, you will want to turn off the electricity or allow the ground to dry before entering. If there is a strong odor such as sewer, you may not want to access due to bacteria and health concerns. Occasionally, a disconnected or damaged soil or waste line dumps waste water or raw sewage into the crawlspace for years. This requires immediate repairs due to possible health hazards.

Water damage due to improper installation of sump pit, drain tile.

Sump Pumps

If there is a sump-pump, check for proper operating

condition and installation. The pump should rest inside a holding tank, lined pit, or even a bucket. If the pump is simply in a dirt hole, this can eventually cause the unit to fail or the hole itself can cave in. Lift the sump-pump float switch and watch and listen for smooth operation. Some internal pumps may have to be activated by adding water to the pit. There may also be drain tile running around the perimeter of the crawlspace and into the sump. Drain tile can help to control water problems by forcing the water into the sump-pump and outside. The drain tile should be below grade, in a trench lined with filter cloth to prevent clogging of the drain tile. The drain tile should rest in a bed of gravel and slope slightly towards the sump pit. If there is water in the sump pump, look inside of the drain tiles and see if they are dry or damp. (See Fig. 2CB)

Properly installed and maintained sump pump.

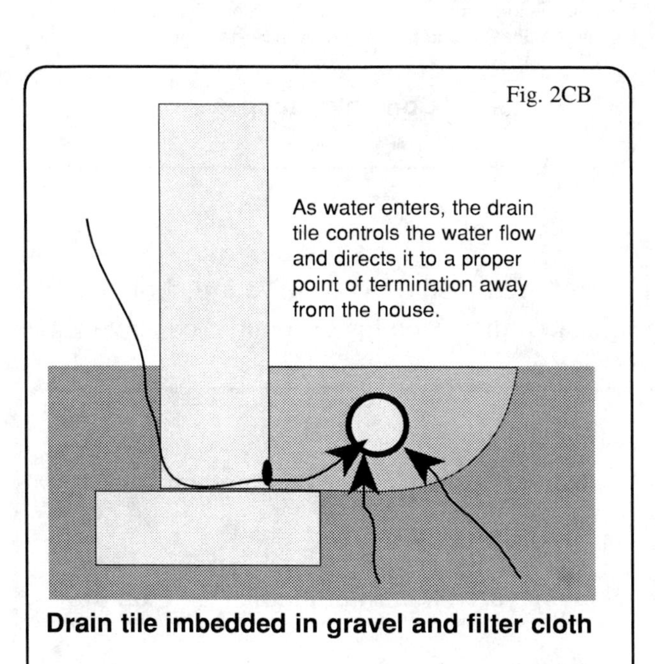

Drain tile imbedded in gravel and filter cloth

The sump pump should have a cover on top. In some areas this may need to be a sealed cover, due to radon. (See "Radon" in the environmental section for more information). If the sump pump is in a remote area, a moisture alarm sensor may be a good idea in case of failure. The sump pump should not be wired to a GFCI which could cause the pump to cut off and overflow. If the condensate drain terminates in the sump pump bacteria could result. Adding peroxide or other disinfectant will help prevent bacteria formation and odors.

Moisture

The floor or grounds should be free of debris and building products. Things such as deteriorating lumber may attract termites, so it should be removed. If there is a plastic vapor barrier on the ground, look for any water collecting on top. This

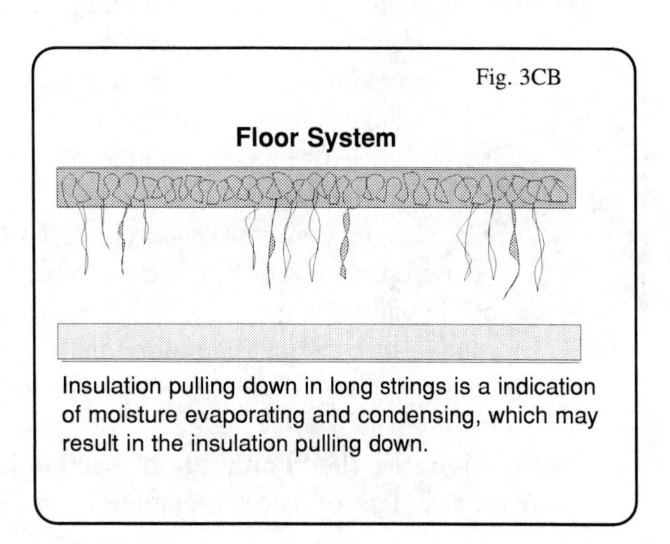

Insulation pulling down in long strings is a indication of moisture evaporating and condensing, which may result in the insulation pulling down.

can cause water evaporation to collect on the floor system above, creating a cycle where water never dries. In some cases water will form droplets on the insulation and floor above, and fall down like rain. Long hanging strings of insulation may indicate a past or present moisture problem; the weight of the condensing moisture causes the insulation to pull down. (See Fig. 3CB) Mildew, fungus and severe rot may also result. In this case, the vapor barrier may have to be removed and replaced with a dry one. If the vapor barrier is partial, moisture can become trapped over the part that does have a vapor barrier and the same situation can occur. A vapor barrier will greatly reduce a moisture problem when installed correctly. It should lap up on the walls, have the seams taped or overlapped, and cover 100% of the crawlspace throughout. A preferred method is to have a bed of smooth washed gravel below the vapor barrier. Some crawlspaces may even have a poured concrete slab.

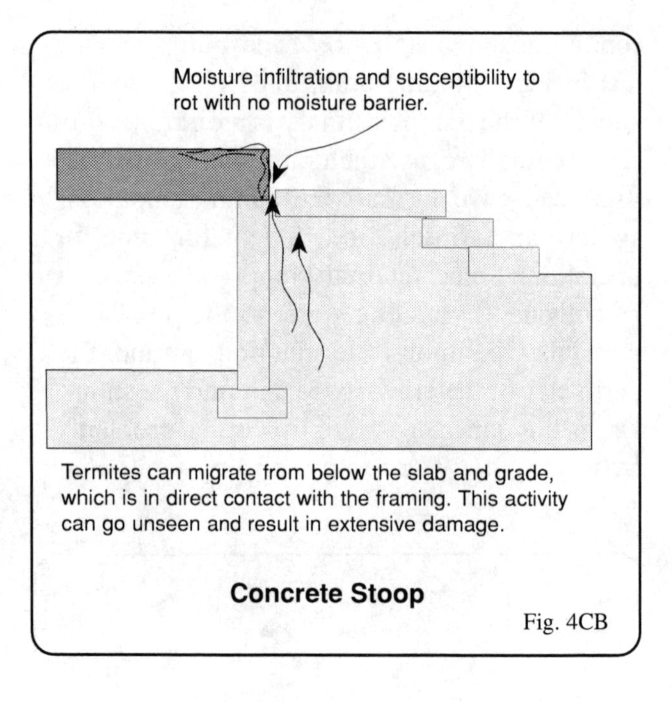

Concrete Stoop

Fig. 4CB

Recommendations on Dampness

To remove damp conditions from crawlspaces, there are several stages recommended.

1. Remove standing water from atop the vapor barrier.
2. Remove debris and grade ground surface.
3. Install perimeter drain tile in a bed of gravel and layered in a filter cloth. The drain tile should terminate in a proper sump pit.
4. Reinstall a proper vapor barrier and lap it up onto the walls of the foundation.
5. Remove grade from contact with crawlspace vents and slope grade away from house.
6. Extend all downspouts and sump pump discharge a minimum of 10' from house.
7. Look for other sources of moisture i.e.:
   - Condensation on plumbing or ductwork.
   - Weep holes above vapor barrier.
   - Inadequate or blocked ventilation.
   - No cross ventilation.
   - Backed up condensation drain on air handler.
   - Improperly run or blocked drain tile.

Once these conditions have been repaired routine monitoring is recommended. All repairs should be conducted by a qualified individual or company.

Foundation (See the "Foundation" section for additional information)

Look for any signs of water or dampness on the foundation walls. Check for efflorescence or

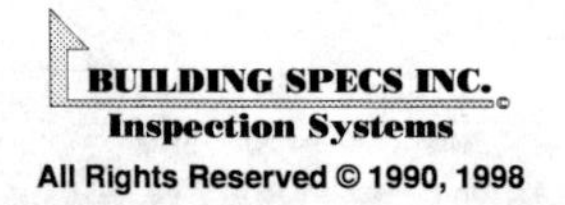

deterioration. Efflorescence is evident when water (hydrostatic pressure) washes salts or minerals from the ground or mortar, out onto the face of concrete block or brick. Its presence is indicated by a white powdery substance or white staining on the surface.

Look for the presence of proper ventilation, preferably cross ventilation. If the vents are adjustable check for proper working condition. Be sure the vents are not below grade; this may allow ground water to enter.

When inspecting the foundation, check for settlement cracks. You may want to note signs of daylight coming through cracks. Check the piers for leaning or settling which could indicate a problem, such as inadequate footings. If there are any wooden or metal supports, inspect for proper footing, plumbness, and proper spacing between the supports. Temporary piers should be noted and they may require further evaluation. Look for cracking to the soil around the pier which may indicate setting. Sight down the beam and look for possible settling at various piers. Excessively damp soil conditions could change the soils bearing capacity and allow the piers to settle.

The top of the foundation should either have a solid cap block, poured solid top course, or a termite shield (usually aluminum). This impedes a direct path for termites through the inside of the block. In many areas, modern construction requires a sill sealer on top of the foundation and a pressure treated sill plate. If the sill is not treated, a close inspection of the sill is in order.

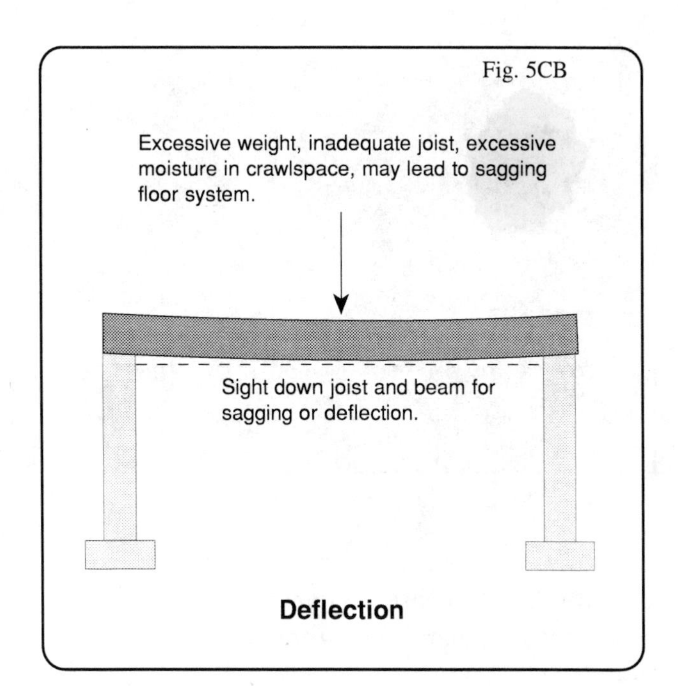

Deflection on a wood beam.

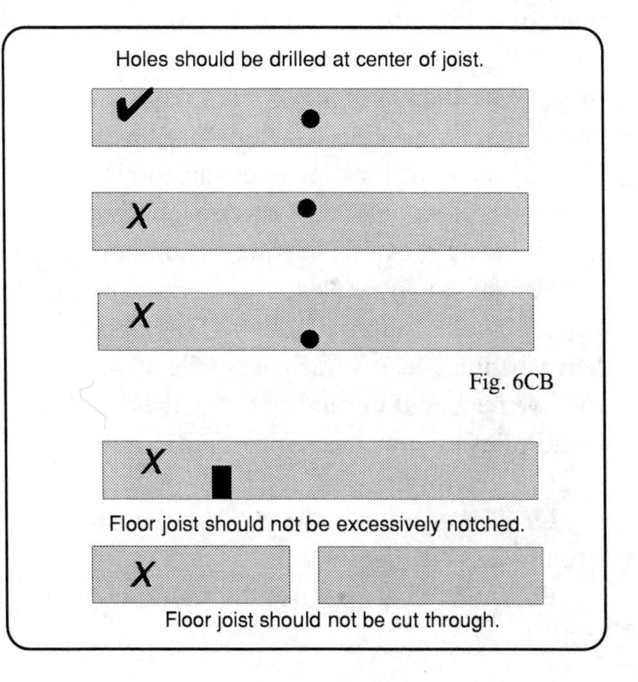

Signs of carpenter bee damage.

Framing

Check for any unsafe frame or structure which could collapse, trap or endanger you. This may be evident by sagging or cracked beams or joists, excessive rot or insect damage, or leaning supports (i.e. piers, wooded post, metal columns (See Fig. 5CB). Probe a representative number of joist and other framing components. Look for fungus, mildew, rot or insect damage. If you have a moisture meter check a representative number of joist for moisture levels. When levels are below 19%, fungus and rot activity should cease. When the levels go above 19%, these may become active again. (See Structure Chapter, "Moisture and Decay"). If the integrity of a framing member has been compromised, structural repairs are likely needed.

When inspecting the framing in the crawlspace you will want to look for water and insect damage as well as proper structural applications. To check for a structurally sound member you can thrust a sharp probe (i.e. awl, screwdriver) into the wood as deeply as possible and even break out some sample areas. Additionally, you can hit the member with a hammer. If the wood breaks apart easily or has a "muted" sound, damage may be present.

Insect Damage

Insect damage may be contained inside wood (i.e. termites) and initially hard to see. For termites, look for mud tunneling up from the foundation or along where the subfloor meets the floor joist; they create tunnels to keep from daylight and to maintain a constant humidity level. If you see plugs or patches uniformly around the foundation or at the slab, this may indicate previous termite treatment.

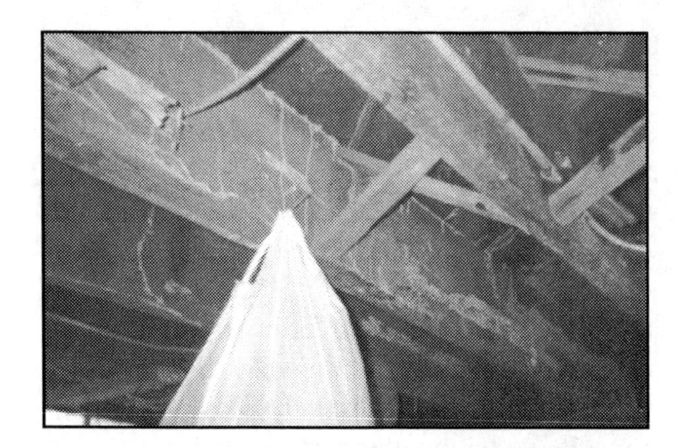
Termites may be identified by mud tubes.

Powder post beetles are evident by small pinholes (approximately 1/16") they leave behind. In several cases, the wood may be peppered throughout with these holes. In some instances, insect damage may be noted by small piles of sawdust lying about. Some termite companies employ the use of specially trained dogs or electronic "sniffers" that indicate the presence of methane gas.

Old wood boar beetles' exit holes are larger (approximately 3/16") than a powder post and usually less frequent.

Carpenter bees (not usually found in crawlspaces) will create an elongated hole in which they tunnel.

Water Damage

Water damage can be identified in several ways. Minor mildew, fungus or molds can lead to wood damage over a period of time. They can be indicators that the conditions for more damage are present. Mildew, fungus or mold can come in many different colors and kinds. Black and white are probably the two most common colors associated with the presence of a destructive fungus. The fungus should be killed and the conditions appropriately altered to prevent reoccurrence. There are chemicals designed to kill various forms of fungus, however bleach and water will kill most forms. More severe dry rot may need structural repair and evaluation. Dry rot can survive from elevated humidity levels alone once established.

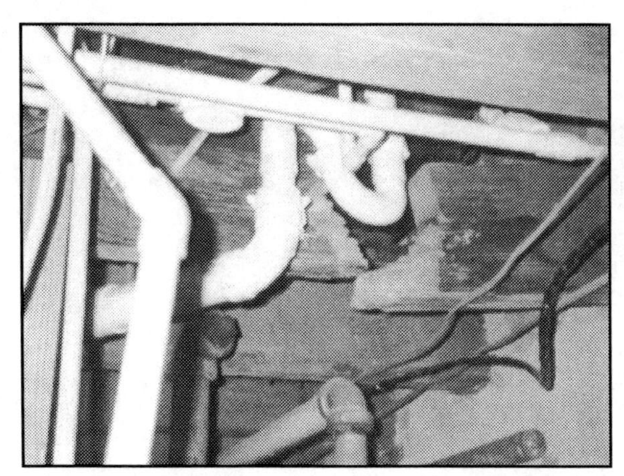

Floor joist integrity compromised for plumbing.

Signs of water damage below leaking tub fixtures.

Floor joists, beams, sills, sheathing and any other wood should be inspected. If there is evidence of damage, it should be determined how much of the member is still sound.

 ***High rot potential areas may be seen at:***

- Plumbing penetrations through floor and walls.
- Low lying outside grade (wood to ground contact).
- Perimeter sills.
- Above standing water locations inside.
- Concrete stoops outside, inspect the framing (if accessible) from behind, by way of the crawlspace for rot.
- Decks or other attachments to the house that have inadequate flashing should be checked for rot.
- Above any condensating water lines or ductwork.

When inspecting the joists and beams note any splitting, checking, cracking (across grain), severe cupping, leaning on bearing points, crushing at the bearing points (joist on the foundation) and signs of sagging or deflection. To check for sagging, simply eye the joist or beam from one end to the other. A string-line can be pulled from one end to the other to get a more exact measurement. A joist that has a point load may show signs of leaning. For example: a joist at a

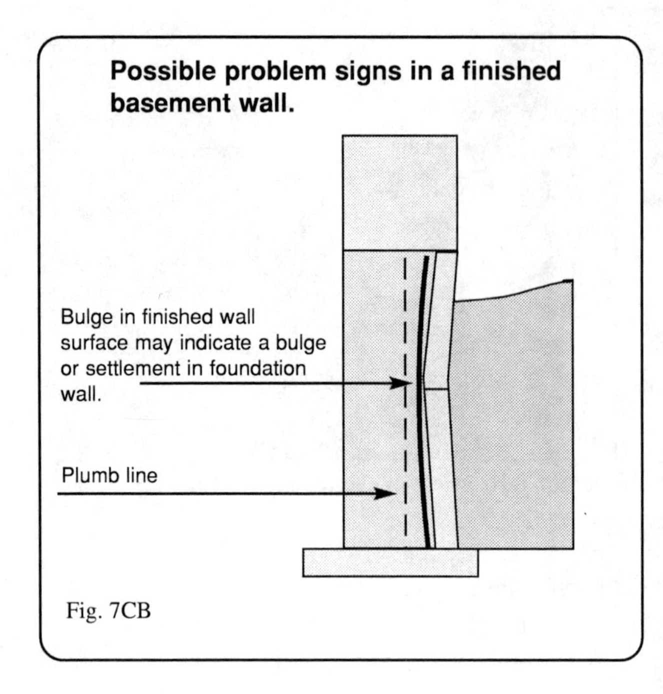

Fig. 7CB

point load may be susceptible to racking when solid blocking is missing for support. By installing bracing between the beam and joist the problem can be alleviated. If the support post is not embedded or securely connected to a footing, racking potential may increase.

Be sure beam and joist dimension, length and spacing are appropriate. If you are in doubt, a BOCA or CABO (national building code guidelines) book will be of assistance.

Any drilled holes should be near the center of the joists or beams. Too large or improperly placed holes may weaken the member. Check for any members that have been cut as to weaken an area. (See Fig. 6CB)

All intersections should have a positive connection, such as joist hangers (metal connector used at joist and ledger). Be sure they are completely and properly nailed. Roofing nails are not considered proper fasteners. They do not have the proper shear-off rating.

 ***Problems to look for : FRAMING***

- Rot, fungus, mildew, insect damage to sill plate, joist(s), beam, subfloor.
- Sagging, cracking, racking, crushing at bearing points to joist(s) or beams.
- No plate or a non-treated sill plate, open top course of block, susceptible to insect or moisture infiltration.
- Elevated moisture or damp conditions.
- Hanging strings of fiberglass insulation, due to evaporation and condensing of moisture.

Other Things to Note

Look for any mechanical or electrical components lying on the ground. This can lead to corrosion, deterioration and premature failure. Electrical lines on the ground can also be hazardous due to possible electrical shock. Condensation drains on Heat/AC units are notorious for flooding. Remember to have them checked when the unit is serviced.

Check the plumbing for leaks and proper hanging. Straps that hold the pipes will sometimes fail; this can cause fatiguing to the pipe. Look for pipe freeze susceptibility. If the floors are insulated and there is no heat in the crawlspace, the pipes may need to be insulated. Inspect the plumbing connections, especially at the floor penetrations. You may have to pull down some insulation. Toilet flanges are a high problem area. If the water pump and tank are in the crawlspace be sure they are not settling in mud and dirt; they should be on a dry stable surface, which will protect them from rust and damage.

Corrosion on steel beams, joist hangers and other metal fasteners should be noted as a symptom of possible dampness. Rust and corrosion should be inhibited in some way, and a professional may be needed. Be sure a qualified individual is used to carry out any repairs.

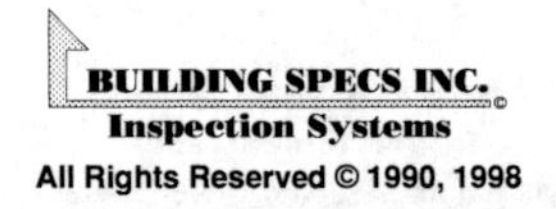

Signs of water seepage in block foundation. Possibly due to standing water in drain tile or improper downspout termination.

Signs of moisture problems:
Mildew or water stains at walls or carpet backing. Look for cupping vinyl tiles or hardwood floor. Water stains at boxes on floor, bottom of stair stringers or other wood in contact with floor.

Fig. 8CB

BASEMENT
(See "Foundations" and "Crawlspaces" for additional information)

Many traits of an unfinished basement are similar to those of crawlspaces. A crawlspace is more inclined to have damp conditions than a basement, but a damp basement is more of a concern, and may be more difficult to repair. Unfortunately, when a basement is finished, a thorough evaluation of the foundation is not possible. Moisture problems or structural cracks may go undetected, until they become extensive enough to show a sign or symptom through the wall.

 ***Leak indicators to look for in a finished basement:***

- Water stains around the perimeter of the floor, carpet backing, walls wicking up, exposed wood bottom plate, at bottom of basement stairs stringers.
- Boxes and personal belonging not directly on floor, resting on pallets, etc.
- Bulge in unfinished wall, which may indicate a bulge in the foundation. (See Fig. 7CB)

If moisture problems are suspected further evaluation may be in order, which may involve an invasive inspection.

 ***Finding the source of your moisture problems:*** (See Fig. 8CB)

- Check for proper grading outside.
- Look for any holes or pocketing next to the foundation.
- Verify all downspout, runoff, and ground water is away from the house.
- Look for any tree roots which could damage the foundation.

See "Foundation" & "Grounds" for more information.

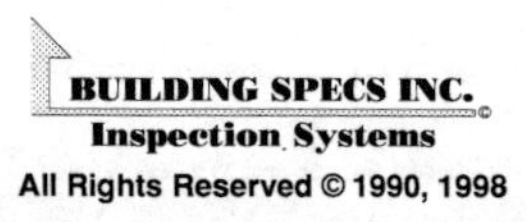

Floor

Most basements will have a concrete floor base, whether finished or not. Cracking in the floor is not uncommon, however the severity of the cracking should be assessed. With a finished basement, a thorough inspection of the concrete floor may not be possible.

Pull carpet back from perimeter --- may expose water stains.

 ***Severe cracking could indicate:***

- A water problem.
- Bad concrete mix.
- The concrete was poured and froze before curing.
- No expansion joint at perimeter.
- No control joint at intersecting rooms.
- Weight applied to the slab without a proper footing, creating a point load for which the slab was not designed.

Some cracking is typical in a slab. The corner near a sump pump even on a newer house commonly cracks. Adjacent rooms where there are two different size slabs and differing expansion coefficients can suffer cracking near the intersection.

Standing water in basement.

Basement entry susceptible to water infiltration. Maintain sealant at threshold and at casing. Keep exterior drain clear from debris.

WATER PROOFING RECOMMENDATIONS

Many times proper grading and water control around the house will take care of minor water infiltration into a basement or crawlspace. The following are some step-by-step recommendations, starting with the least expensive. Once the conditions have been changed, routine monitoring of all of the conditions should be followed.

1. Fill any pockets or holes along the foundation, create a grade away from the house minimum 1/2" per foot.
2. Extend all downspouts and other water discharges a minimum of 10' away from structure.
3. Remove any overgrown vegetation, bushes, trees, or root systems.
4. Seal the interior of the walls with a masonry sealer, fill any holes with a hydraulic cement.
5. Verify all existing drain tiles for sump pumps are clear.

Monitor all conditions.

If the conditions persist:

1. Install a drainfield around the affected areas in the basement, below the slab. A drain tile embedded in gravel gives rising ground water a free path to a sump pump, and to be discharged away from the structure a minimum of 10' feet.
2. There are soil injections available which come with a lifetime warranty. A polymer is injected into the soil around the foundation perimeter which seals any cracks or holes in the foundation.
3. There are some other methods which involve collecting the water inside the wall. Keep in mind that these methods may allow deterioration and water damage to the foundation wall.

For these problems, a professional basement waterproofing company could be consulted as a last resort.

BASEMENT & CRAWLSPACE CHECKLIST

☐Crawl Space / ☐Unfinished Basement

☐*Further Structural Evaluation & Repairs Required, (See Foundation & Structure)*

1] **HAND DUG** ☐No ☐Yes; Total / Partial ☐**Problem;** *Undermined Footing(s)*[1] / Requires Proper Footing(s), Retaining Wall System(s)

2] **ACCESS** ☐Yes ☐N/V ☐N/A ☐**Problem;** Partial / Limited Clearance / Standing Water / Falling Insulation / Obstructed

3] **ACCESS PANEL** ☐Yes ☐**No** ☐N/A ☐**Problem;** Water Entry / Damaged / Stuck / Blocked / Missing / Rusted / Rotted / Inadequate

4] **GROUND** ☐**Dirt** ☐**Gravel** ☐**Mud, Erosion** [☐**Concrete Slab** ☐**Problem**; Cracking / Settling / Hollow Soundings / Amateur]

☐**Problem;** Standing Water / Damp / Needs Grading / [***Remove***; Building Debris / Trash / Fungus / Rotting Wood / Damp Conditions / Pest(s)]

a] **Remove From Ground Contact;** Electrical / Plumbing / Ductwork / Framing / Insulation / HVAC System / Well Tank / Other__________

b] **Vapor Barrier** ☐Yes ☐N/A ☐**Problem;** Missing / Partial / Wet / Required / Improperly Installed / Should Lap up onto Wall / Replace

5] **MOISTURE**[1] ☐N/V ☐Yes ☐**Elevated Moisture Readings**______% & ______% Observed at__________________________

☐**Problem;** Dampness / Standing Water / Efflorescence / Seepage / Water Stains / Metal Corroded / Mildew, Fungus / Rot / Rust Observed

6] **SUMP PUMP** ☐Yes ☐N/V ☐*Could Not Operate Internal Switch* ☐Pit With No Pump ☐Pump with Inadequate Pit

☐**Problem;** *Recommended* / Poor Installation / Pit Failing / Improper Drainage / Not Operational / Drain Tile Improperly Installed / Switch Binds

7] **INSULATION** ☐Yes ☐N/V ☐**Problem;** *Obstructing Inspection* / Falling Down / Stringing Down / Wet / Missing ☐**Recommended**

8] **VENTILATION** ☐Yes ☐N/A ☐**Problem;** None / Inadequate / Below Grade, Water Infiltration / Damaged / Blocked / Loose / Stuck

9] **INTERIOR PIERS / SUPPORTS** ☐Yes ☐N/A ☐**Block** ☐**Brick** ☐**Wood** ☐**Steel** ☐**Other**__________

☐**Problem**[1]; Amteur Work / Leaning / Inadequate Footing / Inadequate Installation / [*Inadequate Supports Under*, Joist, Beam, Stair Landing]

10] **FRAMING** ☐Yes ☐***N/V;*** [Sill Plate / Joist / Subfloor / Beams / Walls / Ceiling / Below Stair Landing / Areas Blocked By Items, etc.]

☐**Problem**[1]; ☐Minor ☐**Severe** ☐**Extensive;** Rot / Insect Damage / Water Damage / Water Stains / Mildew, Fungus / Rusting Joist, Etc.
Cracking / Notched / Drilled / Crushing / Missing Crush Blocks / Recommend Sistering, Bridging / Missing Hangers, Nails / Amateur Work

☐**Observed At;**[1] Sill Plate / Joist / Subfloor / Beams / Walls / Ceiling / Below Stair Landing ☐*Inadequate Shims, Blocking Under Beams*

11] **SUBFLOOR** ☐Yes ☐**Problem;** ☐Minor ☐**Severe;** Rot / Sagging / Water Stains / Water Damage / Old Repairs / Inadequate Repairs

12] **ENTRY EXTERIOR;** ☐Yes ☐N/A ☐**Problem;** Susceptible To Water Infiltration / Water Stains, Damage / Rusting / Drain Clogged

13] **STAIRS** ☐Yes ☐N/A ☐**Problem;** Inadequate Connections / Hazardous / Cracked; Stringer, Treads / Visible Rot / Missing Handrail

•**ADDITIONAL COMMENTS** *[1]Elevated moisture levels in a basement or crawlspace could lead to mildew or rotted wood and possible structural problems. Continuously damp or wet soil could lead to unstable ground conditions, allow piers and/or footings to settle, and could lead to a structural problem.*

____

____

____

____

____

Finished Basement[1]

☐*Further Evaluation Recommended* ☐*Water Proofing Recommended* ☐**Partially Finished**[1]

1] **WALLS** ☐Drywall ☐Wood ☐Paneling ☐Insulation ☐*Freshly Painted, Could Not Properly Evaluate Condition*

☐**Problem;** Water Stains / Damp / Soaked / Damaged / Mildew / Rot / Loose / Amateur Workmanship / Falling Down / Undulation Observed

2] **FLOOR** ☐Carpet ☐Vinyl, Linoleum Tile ☐Ceramic ☐Sheet Goods ☐Wood ☐Concrete*(see above)*

☐**Problem;** Water Stains / Damp / Water Damage / Soaked / Damaged / Mildew / Curling / Loose / Heavy Wear / Cracking / Cupping / Rotted

3] **CEILING** ☐Drywall ☐Drop ☐Wood ☐Paneling ☐Insulation ☐Tiles ☐Other__________

☐**Problem;** Water Stains / Water Damage / Soaked / Damaged / Mildew / Loose / Falling Down / Sagging / Amateur Installation / Undulation

•**ADDITIONAL COMMENTS** *[1]If walls or floors are covered, finished or blocked, a thorough inspection is not possible. If a problem is suspected, the wall coverings should be removed for a complete inspection. Further evaluation and structural repairs recommended by either a licensed contractor, or qualified individual.*

____

____

____

Section Three:

Outside

Grounds

You were warned: routine maintenance means more than just inspecting the actual home. The grounds around the house are very important, and water, trees, roots, bushes, soil and grading are all things which may influence the condition of a house. During your war against deterioration, while patrolling the perimeter of your home, you should keep a sharp eye out for enemies like standing water, damp or wet basements and crawlspaces, cracking or movement in the foundation, and the undermining of footings.

DRAINAGE

The grade around the foundation and the perimeter of the building is frequently overlooked. The grade should not slope back towards the house, and a slight or significant slope away from the house is recommended. Look for signs of erosion or holes in the ground, particularly close to the foundation; this could indicate a pocket which could trap water next to the foundation. If there

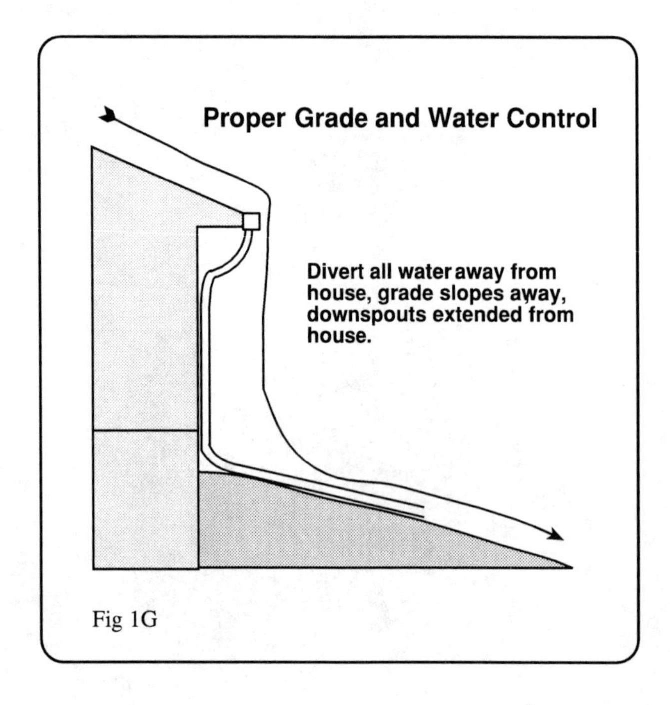

Fig 1G

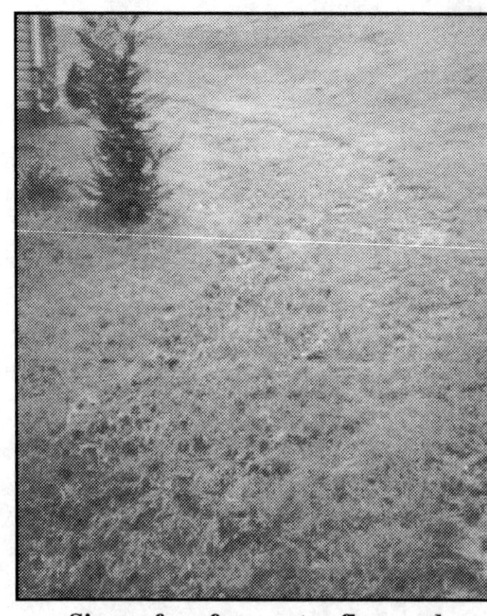

Signs of surface water flow and drainage.

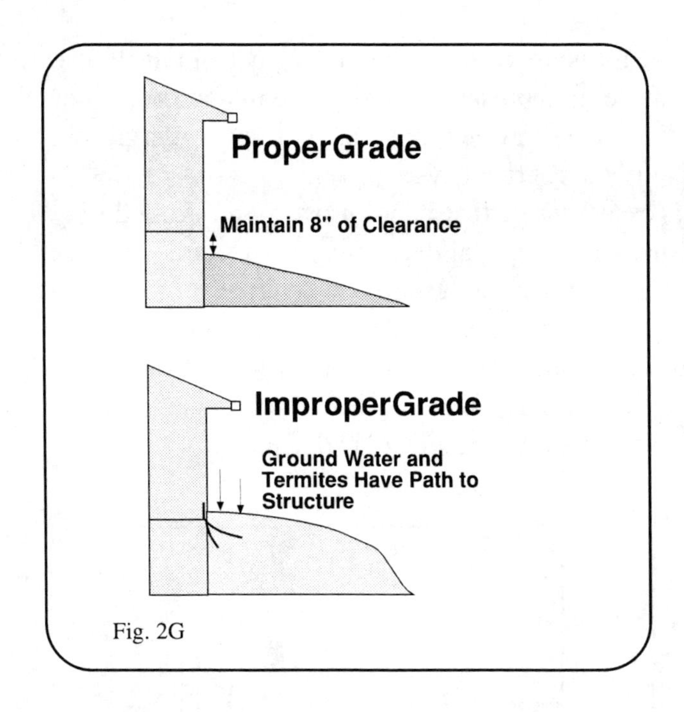

Fig. 2G

is a grading problem, it should be corrected; water problems can deteriorate the footing, foundation, siding, and even affect the framing. Improper grading and water control can lead to settlement cracks in the foundation and even structural failure. All the internal maintenance in the world will not mean much if your whole foundation collapses beneath you.

Many times a damp basement or crawlspace may be corrected by the use of proper drainage and grading. (See Fig 1G).

The grade should be maintained so a minimum distance of 8" is kept from the siding and wall sheathing. The building up of mulch should be discouraged, and existing mulch should be removed and replaced. If not removed, in time the mulch will come in contact with the siding and could promote damp conditions, leading to termite or water infiltration and damage. (See Fig. 2G). If the siding is in contact the conditions should be altered, otherwise, plan on paying for termite treatment.

As the soil around the house becomes saturated with water, a condition known as equilibrium will occur within the soil. This means once the soil surrounding the house cannot hold any more water, water will seek a drier area, creeping towards your home through capillary action. The most likely places for water to go are under a crawlspace or into a foundation wall, or it may just pond on the surface creating run-off. This run-off can cause erosion if not properly controlled.

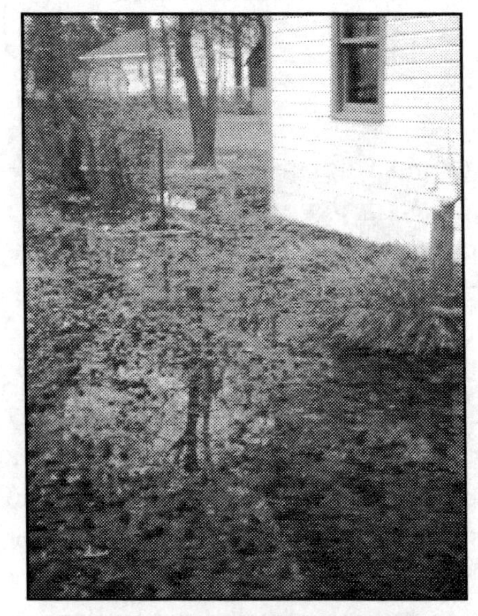

Standing water could adversely affect soil stability and lead to structural settlement.

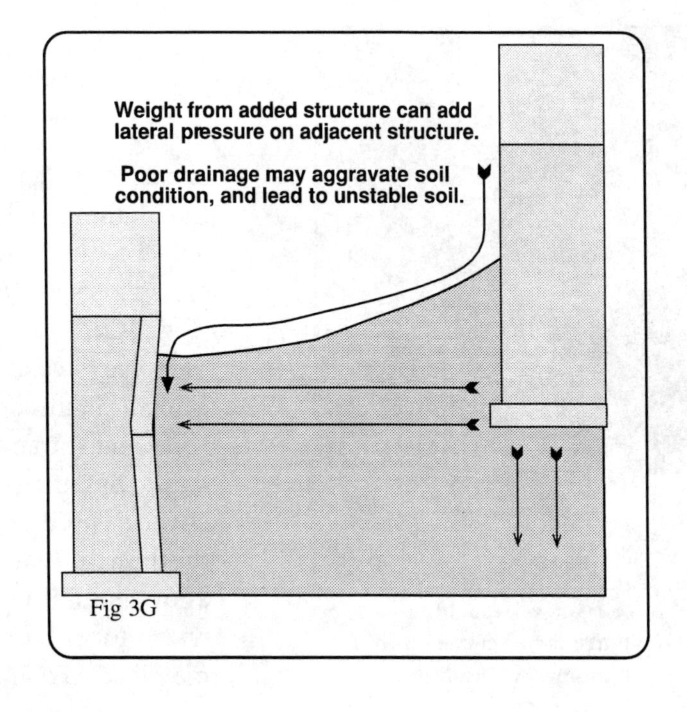

Fig 3G

Connect adjacent downspouts to subterrainian drain tile and extend from house.

Excessive moisture may be diverted away from the house with a swale (a depression in the grade that allows water to flow in a controlled manner), or an underground drainage tile and gravel system.

Excessive water that is not properly controlled may also cause structural failure through a force known as hydrostatic pressure. If ground water freezes the force can create a frostline crack in a foundation (see "Foundation" for more detailed information). The hydrostatic pressure force against the foundation may vary, depending on the soils and adjacent structures.

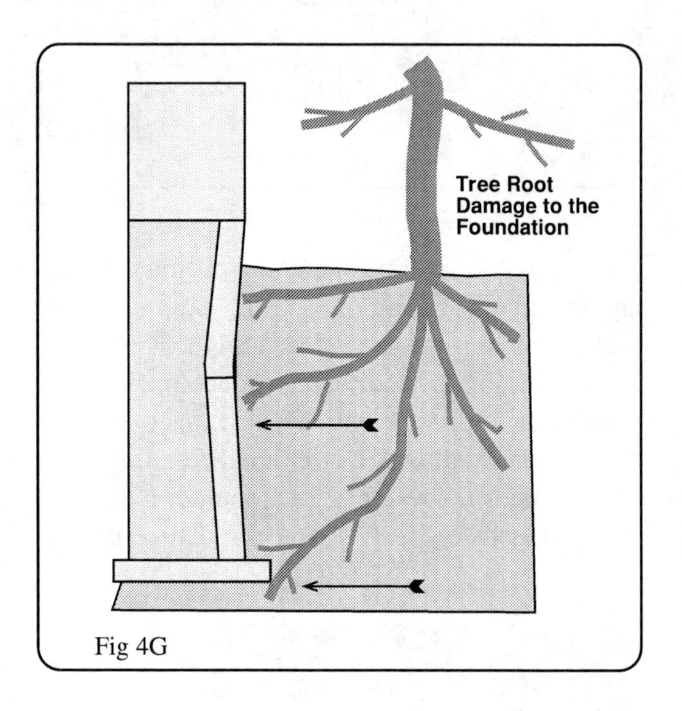

For example, assume your foundation is surrounded by a loose, sandy soil. Ten feet away a hard pan clay vein runs parallel with the sandy vein. The water that is absorbed along the house will not absorb into the hardpan clay as easily as the sandy soil. The clay may become a barrier for the water to push against back towards the foundation. And, regardless of how well-built your home may be, the foundation is more likely to move than the clay. This same principle may apply to a structure as well. (Fig. 3G).

Tree roots encroaching on a structure could cause uplift or damage to foundation.

Any water should be free to move and evaporate. Mulch, landscaping timbers, and retaining walls may sometimes trap water against the building, creating a damp basement or crawlspace. Often a plastic ground cover is used rather than a drainage or filter cloth under the mulch. This will cause the water to stand and not allow natural drainage to occur, as well as not allowing water to evaporate. The trapped water under the cover may lead to damp conditions in a basement or crawlspace and should be removed.

If gutters are installed, check to see if they are leaking or overflowing. Evidence of failing gutters may be determined by an erosion line on the ground beneath the gutters. The main purpose for gutters is to collect and properly control the water shed away from the structure. The downspout should have splash guards to help break the water force to the ground. It is recommended to install downspout extensions or drain tile to get the water away from the house. The drain tile may be terminated in a french drain --- a gravel filled pit lined with filter cloth --- 10 or more feet from the foundation. Sump pumps and condensation drains should be diverted away as well.

MOTHER NATURE MOVING IN ON YOUR TERRITORY

Roots from trees and bushes can cause problems with the footing and foundation, and should be cut back if they are in contact with the house. (Fig. 4G). Tree roots in particular can push on the foundation and lead to structural failure or water infiltration. It is recommended to keep tree roots back from the house.

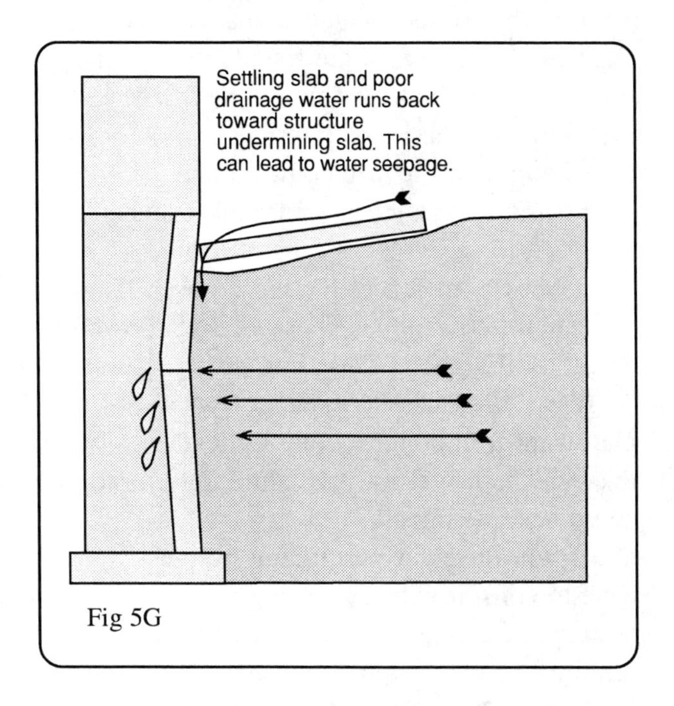

Fig 5G

Controlling plant life around the house will help air to circulate and help the ground around the house to dry. Branches rubbing the house may damage siding, windows and roofing materials. Encroaching branches should be cut back from power lines, cable or telephone wires. Check to see if your local utility company will take care of this for you at no charge. **Do not attempt to cut branches back from power lines yourself; it can easily lead to a shock hazards**. Ivy can cause great damage by getting under trim and siding and thereby causing water problems.

Grade at window may allow water or insect infiltration or damage to occur. Recommend installation of window well and plastic cover.

MAINTAINING RETAINING WALLS

Inspect any retaining walls for failing, rotting or leaning problems. Retaining walls need support that are referred to as "dead men." They are designed to hold back, for front bracing, or as supports that are driven or buried deep into the ground. Without the dead men, the wall may fail. Retaining walls need a place for water pressure to escape. Water pressure that builds behind a wall can cause premature failures, and even low walls need a space from which the water can drain. The pressure will cause mortar joints in block and brick to erode, and block and wooden walls can lean from the force and eventually fail. Filter cloth is recommended behind the wall to prevent dirt from washing through the drains, known as "weep holes."

SIDEWALKS, DRIVEWAYS, PATIOS & FENCES

Look for any signs of settling at sidewalks, driveways or patios. This could indicate a drainage problem or undermining of the slab. Problems can be verified by using a dead blow hammer and listening for sounding and a change in density sounds. There could be pockets under the slab trapping water against the foundation. A slab or sidewalk should slope away from structure, and any gap along the house should be filled with a polyurethane caulk. Potential gap areas include:

- Driveway to apron (sloping section before the driveway) connection.
- Foundation to slab connection.
- Basement stairs and sidewalk connection.

Check your fences for loose posts. If the post are wood, check with a sharp probe to see if they are rotted at or below grade. Sight down the length to make sure no posts are severely warped or out of alignment. This is more for aesthetics than anything.

You will also want to check your driveway. Concrete can settle and crack, and if it does, you may want to monitor its rate of deterioration. Asphalt driveways should be sealed as needed, as this will help prevent premature deterioration. Gravel and dirt should be checked for ruts that could become impassable.

If you have severe grading problems you may have to consult with a civil engineer or a landscape architect. In some areas a grading permit might be required, so check your local codes. An experienced landscaper can devise functional designs to take care of grading problems.

Signs of gutters overspilling. Water infiltration into crawlspace vent.

GROUNDS CHECKLIST

Grounds ☐Appears Adequate ☐*Proper Grading and Surface Water Control Required*[1] ☐Not Visible

1] **CONDITIONS** ☐**Dry** ☐**Damp / Standing Water**[1] **/ Mud / Snow ____" / Patches of Snow / Ice / Frozen / Dew / Covered With Leaves**[2]

2] **DRAINAGE NEXT TO HOUSE** ☐Adequate ☐**Problem**[1]; Improper, Negative Grade / Mulch, Grade in Contact or Close to Siding[1]
☐Near Level ☐Moderate Slope ☐Steep Slope ☐***Graded Back To house***[1] *Water damage and/or infiltration possible to house.*

3] **EROSION** ☐N/V ☐Minor ☐**Severe** ☐Soil Cracked / Mud / Holes / Standing Water / Structure Losing Soil Bearing
Front / Back / Left Side / Right Side ***Observed at***; Down Spout(s) / Sump Pump, Condensate Discharge / Foundation Perimeter
☐***Recommend***[1]; Proper Grading / Filter Cloth / Ground Cover / Drain Tile / Splash Guards / Swale / Retaining Wall / Civil Engineer

4] **TREES, BUSHES** ☐N/A ☐Ok ☐Minor ☐**Severe** ☐***Damaging***; Siding / Foundation / Roofing Material / Wires / Other
☐***Recommend***[1]; Trimming Branches / Remove Roots, Stump(s) / Improving Restricted Air Circulation / Remove Ivy and Other Vegetation

5] **CONCRETE, BRICK, WOOD STOOP / STAIRS / BASEMENT STAIRS / SIDEWALKS** ☐N/A ☐Ok ☐**Repair / Replace As Needed**
☐**Problem;** Settling / Cracked / Water Damage / Rotted / Hazardous / Hollow Soundings / Amateur Workmanship / Loose Bricks, Concrete

6] **RAILING** ☐N/A ☐Yes ☐**Problem;** Rusted / Rusted Off At Base / Damaged / Failing / Loose / Missing / Recommended / Rotted

7] **DRIVEWAY** ☐N/A ☐Yes ☐**Concrete** ☐**Asphalt** ☐**Stamped Concrete** ☐**Brick** ☐**Gravel** ☐**Dirt**
☐**Problem;** Cracked / Settling / Water Damage / Ruts / Poor Condition / Pot Holes ☐***Recommend Sealing / Resurfacing / Regrading***

8] **FENCE** *(Limited to Contact With House)* ☐N/A ☐**Chain Link** ☐**Board on Board** ☐**Picket** ☐**Split Rail** ☐**Wrought Iron**
☐**Problem;** Water, Insect Damage / Rot / Loose / Rusted / Damaged / Gate Sagging / Gate Binding / Poor Condition / Falling Apart

9] **RETAINING / BASEMENT ENTRY WALL** ☐N/A ☐Yes ☐Block ☐Wood ☐Concrete ☐Metal ☐Stone ☐Brick
☐**Problem;** Recommended / Leaning / Cracking / No Water Relief / Severely Failing / Rotting / Undermined Footing / Hillside Failing

10] **FOUNDATION VENTS / WINDOWS** ☐***Susceptible to Water Infiltration*** ☐**Recommend Wells / Covers** ☐**Damaged, Blocked**

•**ADDITIONAL COMMENTS** *[1]Proper grading and water control is essential to prevent water, and/or insect damage to siding and structure. [2]Grading and exterior cannot be evaluated if covered, a reinspection is recommended. Constantly damp soil could lead to foundation settlement, cracking and failure.*

Exterior

The exterior of the building is essentially a barrier between the elements and the superstructure, the building's components and the occupants. If any of the exterior components fail or deteriorate, water infiltration and potential damage to visible and/or concealed interior sections of the building may occur. The exterior of a building may be made from various combined materials, and proper application (according to manufacturer's specifications) of these products is important to insure adequate protection to the building.

SIDING

There are many types, styles, and materials used to side a house. They all have different applications and problems associated with them. However, they all serve the same purpose: to keep the elements out and protect the building's components. Some materials are virtually maintenance free, while others may require frequent upkeep.

Most sidings require some basic knowledge of installation. They all have their own little idiosyncrasies and tricks for proper installation. Amateur or poor installation can mean the difference in not only aesthetics, but functional problems as well. The siding's function is to protect the building from the elements such as wind, rain and sun. Improper application or installation could allow for premature failure.

Some sidings may have been applied over an existing siding, creating a build-up of multiple layers over the years. Look under the bottom edge of the siding for signs of previous layers where previous damage may be concealed.

You might also have a porch that once was enclosed. A problem to look for in this case is the tongue and groove flooring material protruding past the siding. This situation leaves the floor susceptible to water infiltration, and rot may occur to the flooring and the wall. The flooring should be cut back and the siding adjusted so that it covers the entire wall.

You can also sight down the rows of a horizontal siding material to identify settlement. Look for sagging or bulging to the wall, inspect the butt ends of the siding, and if it appears to be sagging, look for a tighter connection at the top and a gap at the bottom. If you notice some of these signs it may not necessarily be time to panic --- in some cases this might only imply poor workmanship.

If a porch on a concrete slab has been added, often the slab will stick out past the wall. This gives the water a place to seep into the wall and can lead to rot and/or insect damage. Cutting the

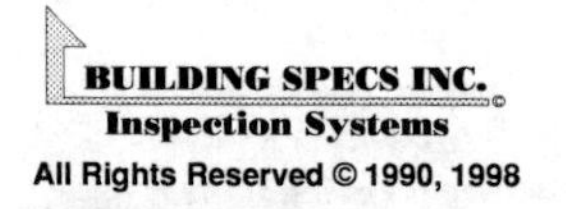

concrete slab is not necessarily the correct repair and can be time consuming, expensive, and could also weaken the bearing point below the wall. If there is enough of a bearing point, the slab may be trimmed back and sided over. There are also two other correction procedures: The first is to install flashing that runs up behind the siding, and is bent down over the face of the slab. This forces the water away from the structure. The second is to either move the wall, or to build the wall out to the edge of the slab. This way the siding can come over the edge of the concrete and create a proper seal.

Fig. 1E

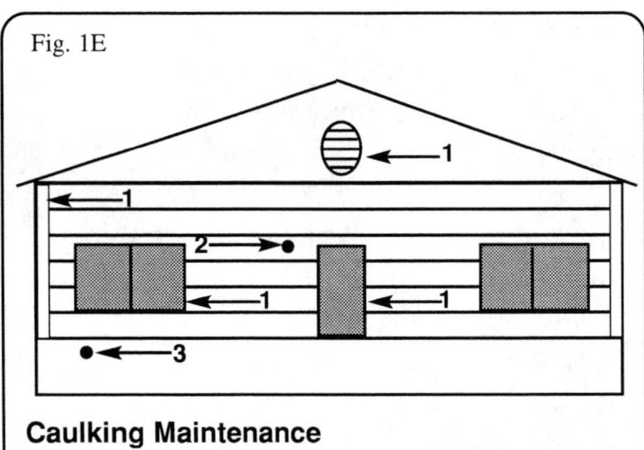

Caulking Maintenance

1. Around all exterior doors, windows, vents, trim, etc.
2. Light fixtures and receptacles.
3. All wall and foundation penetrations including: hose bibs, heating lines, electric lines, etc.

These are all high potential moisture problems with all siding materials.

Vinyl

Vinyl is a very durable and virtually maintenance free siding material. It comes in many different styles including some that even emulate a wood grain. These are some styles:

- Clapboard (double, triple, single, etc.)
- Dutchlap
- Beaded
- Vertical
- Faux cedar shingle panels

Vinyl sections work as interlocking panels and should be nailed over a solid flat sheathing such as plywood or oriented strand board. Unfortunately, they are often nailed directly over old siding products such as asbestos or cedar shingles, and this can create an irregular finish.

Panels should be nailed with aluminum siding nails. Galvanized nails may rust and cause rust streaks on the face of the siding. Nails should be fastened directly into the studs or solid framing and not just the sheathing, otherwise the nails may pull out of the sheathing and leave the siding loose. This occurs due to heat buildup in the wall and expansion and contraction over time. The nails should not be nailed too tightly, so the panels have room to expand and contract. Manufacturers recommend 1/32" to 1/16" between the nail head and the siding.

Definitions...

Soffit: Horizontal overhang adjacent to gutters.

The panels should overlap about 1-1/2" at the ends. If the overlap is not sufficient, water could be drawn in and create yet a another moisture problem. Separate the siding at the overlap randomly to reveal the amount of overlap. If caulking is noted at any overlaps, it may be from a previous leak.

There should be a section --- known as a trim tab --- below long horizontal stretches, such as a soffit or window. Throughout this section, the siding its perforated with a special tool and is then snapped into the trim tab. It is common to see the siding tucked into a j-channel and tacked with a small nail, but with this method the siding often buckles or falls down. One trick for repair is to pinch the bottom of the siding track with bullnose pliers. This creates a friction point that will not pop loose.

Look at your siding's horizontal runs for straight lines. If you see any dark areas under the horizontal connections, the siding may not be completely interlocked. Grab the siding where it interlocks and pull out on the panel. Do this in a few areas, especially along the bottom row. If the panel is loose or has excessive play it could work loose or be blown off in high winds.

If you ever have to do an invasive inspection, a zip tool is available that will help you to unlock and engage the siding.

Keep mulch and grade from contact with siding. Insect and water damage could result.

Vinyl is fairly impervious to damage, though hail can be devastating when it causes severe cracks. Minor denting is usually the only outcome of a hail storm, and that is merely an aesthetic nuisance.

If you notice buckling throughout your siding, keep in mind that older vinyl sidings used a styrofoam backing imbedded on the back of the siding panels for added insulation. Sometimes this causes the siding to buckle due to excessive heat build-up.

The windows and trim may be wrapped with j-channel (vinyl or aluminum channel encasing ends of siding) or butted with the siding. In any case, make sure any areas susceptible to water or air infiltration are caulked and sealed. These areas need to be maintained regularly, usually once a year. Higher grades of caulking such as polyurethane may decrease the frequency of repair. (See Fig. 1E).

When inspecting vinyl siding, look for proper installation methods and neat work, as evidenced by tight fits, neat trimming, all sections wrapped appropriately, straight runs, and proper use of special trimming sections. The overlaps should be running all the same direction so the majority of seams disappear. The siding should also overlap, so that in windy areas the wind cannot get behind the siding and pull it from the building. A lack of these features is not necessarily a problem, but may give you a clue as to the competency or quality of the installer. Poor workmanship might imply the greater chance of potential problems.

Trim bushes and trees back from siding to allow for air circulation and to prevent damage to unfinished surfaces.

✔ ***Maintenance may include:***

- Keeping trees and bushes trimmed back so they will not rub and damage the siding.

• Rinsing off when dirt accumulates.
• Maintaining any areas that require caulking: around windows, doors, wood trim, wall penetrations, etc. (See Fig. 1E).
• Repairing any loose or damaged sections.
• Keeping mulch and grade several inches from contact with siding and trims.

Aluminum

Aluminum siding is similar in installation and maintenance as vinyl, but it is more prone to denting and other cosmetic damages. It is also more susceptible to the interlocking strip becoming disengaged. Aluminum expands and contracts more, and if the lip is not properly engaged, or the panels are nailed too loosely, the result may be loose siding. This is not as simple to re-engage as vinyl, and may require some ingenuity. For example: you could try pop riveting the bottom of the loose section of siding to the section below, or you could apply construction adhesive behind the loose panel, at the sheathing and interlocking strip. This may bond the panel in place to the wall sheathing. Removing and re-applying the siding is a final option.

Aluminum also makes a popping or crackling sound as it heats and cools. This again is due to the amount of expansion and contraction to which it is subjected.

Prevent any electrical wiring from using the siding as a ground and becoming a potential electrical shock hazard. When electrical wiring passes through the aluminum siding, inspect the wiring area for the possibility of the siding cutting through the sheathing. Some people consider it advisable to ground the siding to a rod.

Another trait aluminum siding exhibits is chalking. However, there are paints available for aluminum siding if needed.

Maintenance includes the same techniques used on vinyl.

Wood

Wood has been used as a siding for a very long time. Some woods are more impervious than others to the elements, decay, rot and insect damage, including redwood, cedar, cypress, treated, mahogany, teak, and white oak.

Wood siding also comes in many styles including:

• Clapboard (which can be installed with a range of exposures)
• Dutchlap, German, Novelty
• Beaded
• Tongue and Groove
• Vertical
• Board and Batten
• Shingles
• Ripped Logs (on cabins)
• Logs
• Grooved plywood

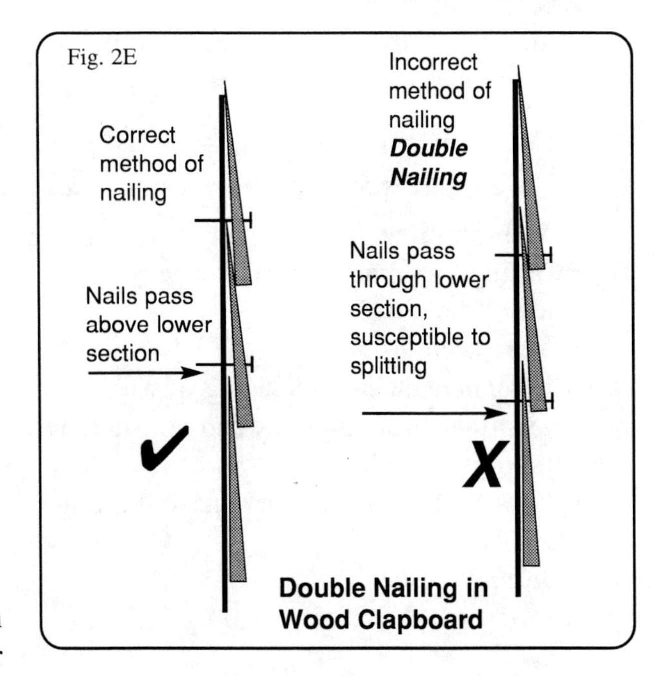

Most wood sidings have certain installation requirement, including proper nailing techniques. When horizontal clapboards are nailed, the nail in the upper board should not penetrate the board it is overlapping. The nail should be about 1/8" above the top of the lower piece, because if the nail penetrates both pieces of siding, splitting can occur. As the upper section becomes wet it swells, and at the same time the lower board is wet, swelling except where it is overlapped. When a nail pierces this area, and the two boards expand in different areas, splitting at the nail line results. (See Fig. 2E)

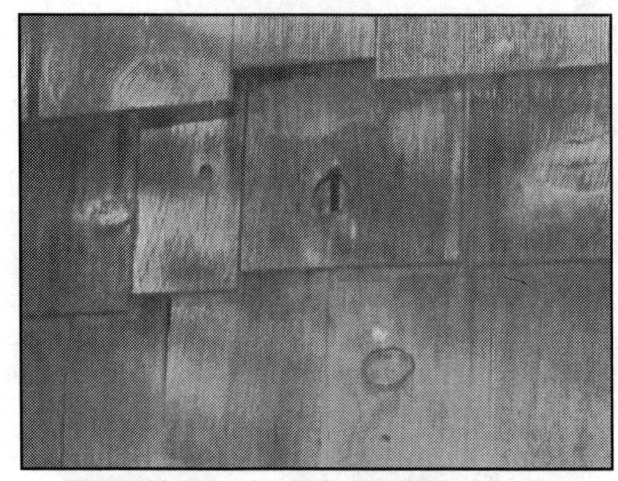

Wood shingle s should be installed over wood sheathing or slats. Loose knot holes may allow for water infiltration.

The nails should penetrate the frame of the house, not just the sheathing. Stainless or aluminum nails are preferable, since galvanized are prone to streaking rust stains on the face. The siding should be fastened with at least an 8d ring shank or spiral nail, with a head. Trim nails tend to pull through the siding and not hold as well.

The siding must be nailed over the appropriate sheathing. Incompatible sheathing examples include:

1. Wood shingles nailed over thermal-ply or impregnated homosote (brown board).
2. Vertical siding nailed without adequate horizontal blocking.

Either of these or other applications may be viewed from an attic gable wall (where applicable). An unfinished garage wall may also show what type of sheathing is present. You may also try looking under the lip of the bottom row of siding for the sheathing. If this is not possible for you, try pulling on some sections of siding to see if they are secure.

With many of these sidings, back priming is highly recommended, as is sealing any cut edges. The back primer may be a clear sealer, stain or paint.

 ***Problems to look for with wood siding include:***

- Double nailing.
- Curling, especially on the southern side of the house.
- Loose sections.
- Inadequate nailing, including missed nails (into sheathing only and not framing).
- Improper sheathing.
- Seams not breaking over stud of framing member.
- Ground contact.
- Failing sealant.
- No caulking.
- Trim applied over horizontal clapboard (and similar) and not butting into the trim.
- No drip edge over doors, windows, and other applications which may allow water to infiltrate.
- Gaps at siding butt joints which could allow moisture infiltration.

 *Maintenance may include:*

- Maintaining sealer on surface of siding, which may include clear sealer, stain or paint.
- Caulking around trim, windows, and all wall penetrations.
- Repairing any loose, cracked, rotted or damaged sections.
- Keeping mulch and dirt from contacting or becoming to close to the wood.
- Trimming back trees and bushes.

Wooden window sills are subject to rot at endgrains.

Pressboard, Masonite, Composite Board

Pressboard, Masonite, and Composite board sidings have evolved since first introduced. These boards were once highly prone to delamination and deterioration, and even now, keep in mind that some are made with better glues and sealers. Louisiana Pacific (L.P.) siding, commonly found on the west coast, has the most notorious failure story, and there is a multi-million dollar lawsuit in the works involving siding installed from 1985-1995. Prolonged damp conditions have caused the product to fail. (For further information on the lawsuit call 800-245-2722.)

Pressboard, Masonite, and Composite boards are available in wood embossed finishes, and with the price of lumber continually rising, these sidings will become more predominant in the industry. These boards still need to have the ends sealed when cut and have the sealers and caulking maintained.

 *Problems to look for may include:*

- Swelling at the joints, due to moisture penetration.
- Deterioration, delamination, rot and failure due to moisture penetration.
- Lack of proper caulks and sealants.

See "Wood Siding" section for more information.

Maintenance may include the same procedures described in the wood siding sections.

Brick

Brick may be used as a veneer over wood-framed walls or masonry walls. This veneer should rest on a solid base. This base may include:

- Extended footing.
- Stepped out block, 8"x8"x16" on top of a 12"x12"x16" concrete block.
- Steel or concrete lintel.

Any of these may be considered adequate for support of the brick veneer. The brick veneer should have metal straps secured back to the wall and imbedded in the mortar joint. If these straps are missing or failing, the brick veneer could fail and fall away from the house.

Brick veneer should also have weep holes across the bottom and at least one horizontal row per story. This will allow any water trapped between the veneer and wall to escape. There should be flashing nailed to the house and laid in the mortar at the weep hole level. This creates a barrier

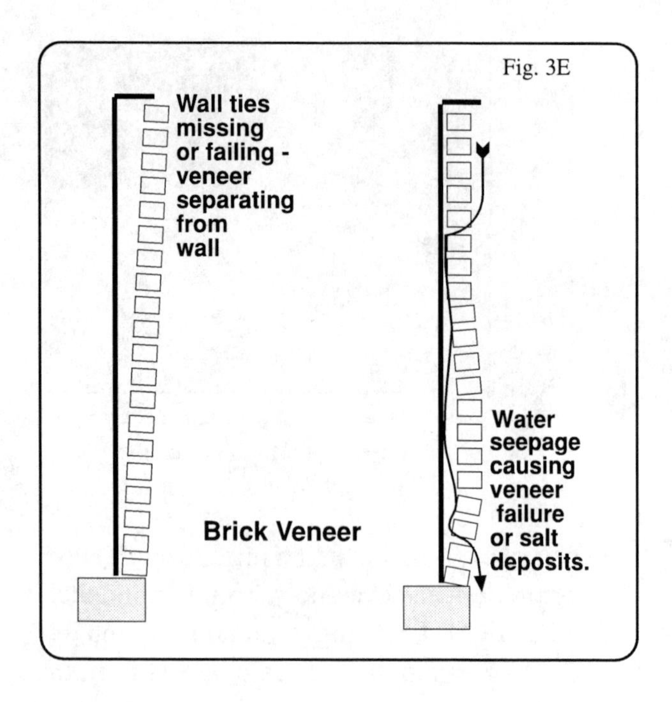

for the water to be forced out and away from the structure, to catch any moisture, and to give it a path out from the cavity. Excessive moisture in this wall cavity could deteriorate the metal tie straps, cause water damage to the wall, and cause the mortar or brick veneer to fail. Efflorescence and spalling could be signs of moisture infiltration (See Fig. 3E).

Inspect the visible outside conditions of the mortar and brick itself. With your fingernail and a sharp probe, check the integrity of the surfaces. Older bricks may be failing, evident by crumbling and/or deterioration. Check to see if the mortar has eroded or is failing. When the face of the brick cracks or peels off in a layer, this is referred to as spalling. Spalling occurs when water seeps into the brick, freezes, and causes the brick to crack. If there are weep holes, check to see if they are cleared from debris.

Pound on the brick veneer with the side of your fist and push in areas, though not so hard that you damage your hand. Look for any movement, loose or failing sections that may be inadequately secured. Inspect the condition of all the window sills, looking for loose bricks. Also check to see if the sill has an appropriate pitch to shed away water. If the sill is flat, or is poorly sloping, water could backup and infiltrate the house or wall. This can also result in damage to the window unit.

If there are any arches, inspect the integrity of the arch and the mortar closely. Bad mortar or poor construction could cause an arch to fail, and a brick to the head is never a pleasant surprise for anyone.

Brick window sills should slope away from windows and be caulked.

Stone Veneer: See "Brick."

Stucco

Stucco is a very durable weather resistant material used extensively in the south, but gaining popularity in the northern regions. Stucco holds up well against the elements and is wind resistant. Moisture infiltration is the main problem with stucco, sometimes causing extensive and costly damage and repairs.

Stucco may be applied over masonry, wood, or exterior drywall. When stucco is applied over wood, a moisture barrier such as a layer of tar paper is required. Otherwise the moisture would be prematurely drawn from the cement when drying, causing improper curing. Also, the moisture barrier prevents any water that penetrates the stucco over the years from damaging the wood framed wall.

Stucco is applied over a wire mesh lathing that is fastened to the wall. Next a cement scratch base is applied over the wire, followed by the finish coats, which may have color impregnated throughout the mix. There are additives and polymers that some succoes use to add to their durability. Polymer-based stuccoes are in some cases applied directly over older uneven siding, and some polymer stucco finishes are applied over a plastic mesh virtually impervious to moisture, air infiltration, and moisture release. If moisture does creep behind this finish, extensive damage may occur.

When inspecting stucco look for loose, cracking or crumbing sections; these could be due to water infiltration somewhere in the wall. Examine the stucco for areas that may be "pocketing" or delaminating. Tap on the wall and push on sections to look for spongy or loose sections. Look for any water damage or stains. Water damage may cause crumbling of the stucco, and moisture may be detected with a pinned conductive moisture meter or a pinless meter which can detect moisture behind a surface. It is essential to determine if any moisture is present in these areas, for any moisture infiltration could lead to the stucco failing as well as extensive structural damage to the building.

 ***Maintenance may include:***

- Repairing any damaged or loose sections.
- Maintaining sealers where needed, including caulking where water could get behind stucco and cause the bond to the wall to fail and delaminate.

Cultured Concrete

This is a cement product which emulates natural stone, applied like a veneered finish. It may be identified by scratching the surface or looking for surface cracking. Look for any loose sections or failing material. There is no recommended maintenance.

Shingles, Asbestos, Cement Board

These shingles from the 1960's and earlier are still used, usually for matching additions and repairs. This type of siding is also often sided-over with vinyl, aluminum or other siding products. When inspecting these, look for any loose, cracked, or damaged shingles.

Often these shingles may contain asbestos. Some companies used asbestos fibers until 1990, but have since started using eucalyptus fibers. If the shingles are to be removed, special precaution needs to be taken. There are professional remediation companies that specialize in the removal of asbestos. If these shingles are left in place there is little chance of contamination; it is just in moving them that more airborne contaminates may result.

See"Environmental" section for more information.

 ***Maintenance may include:***

- Repairing or replacing any loose or damaged sections.
- Caulking at trim, windows or other connections.

• Extra wide gaps between shingles may need caulking at seams.
• Maintaining paint.

Asphalt

These shingles were commonly used in the 1950's and often used as a second layer of siding. Asphalt is seen in some styles resembling brick, stone, wood grain and other materials. This type of siding is common in inner cities and also rural areas. It was an inexpensive way to protect buildings.

As asphalt siding ages, it begins to crack and crumble. Once the asphalt begins to deteriorate, replacement with a conventional siding product is recommended. There is no recommended maintenance for this material.

Steel

Steel is a very durable siding as long as the finish is intact. When the finish is damaged, rust can take hold and cause deterioration and aesthetic problems. Keep in mind to look for any electrical wires that could be potential hazards when in contact with the steel.

Maintenance is the same as care for vinyl and aluminum.

TRIM

Trim may be used as a finish detail to accentuate a component, or it can be of functional use, such as a place for siding to properly terminate. In either case, trim is also an area for moisture to infiltrate.

Trim can be made from:

• Wood
• Aluminum clad wood
• Vinyl
• Aluminum

Whatever kind of trim is used, proper installation is mandatory, or water penetration is likely.

There are two common methods through which wood trim is used with wood clapboard; a good way and a bad way. The bad way is to nail the trim over top of the siding, which does not seal the ends of the siding properly, leaving water and insects to enter the gap between the trim and siding. This situation particularly appeals to bees, who find it an ideal area to nest. The preferred method is to butt the siding against the trim, then caulk the joint. This gives a tight seal to the siding and may save you some summer stings. The trim should be thicker than the siding. For example: siding that is 1/2" thick should butt against a piece of trim that is at least 3/4". If the trim is too thin, the caulking bead will protrude past the trim, and be prone to failure.

With wood trim, maintaining a tight seal with caulk around the windows and doors is imperative. Wicking from the end grain can cause severe water damage. If there are any vertical runs that require a mitered joint, the connection should be so the upper board overlaps in order to force water away. The same would apply to miters at the rake boards.

With vinyl and aluminum, the trim is used to seal the ends of the siding as well as help prevent water infiltration around windows, doors, and other areas. This is accomplished with the use of

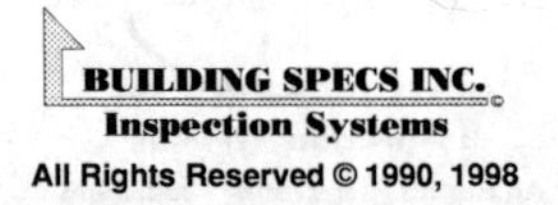

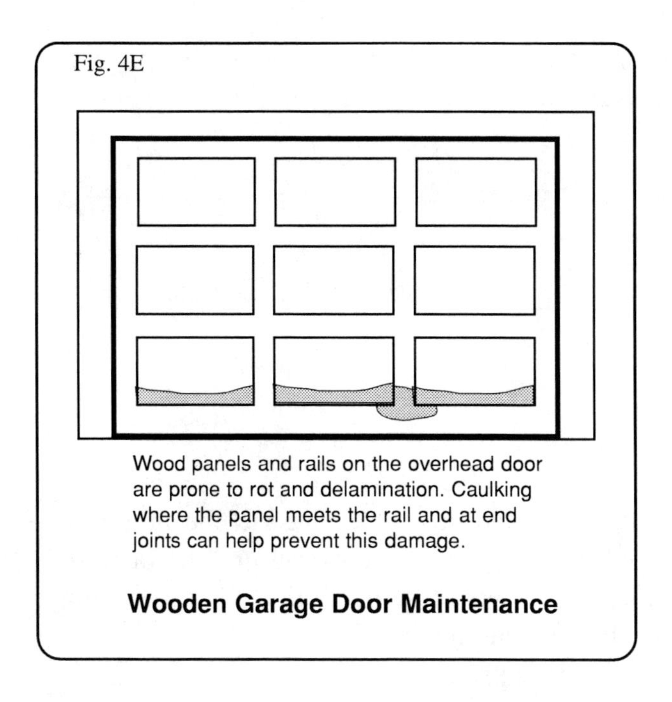

assorted trims, which must always be overlapped so water is shed and cannot get behind the lower section.

Trim Types:

- *J-channel:* used around windows doors, under soffits, rakes, and other areas that need a termination point for the siding.
- *Trim tab:* Recommended but not always used, for the top of a run of siding to lock into. This is used under windows, soffits and other long horizontal expanses. When trim tab is not used, longer sections of siding are prone to working loose.
- *Inside and outside corners:* Used to terminate ends of siding.
- *F-channel:* used for vinyl or aluminum soffit to lock into.
- *Aluminum fascia:* used to encase wood trim, to make it maintenance free.

There are dozens of other decorative trims available but these are the essential components, which help to make the building tight.

Wood trim may also be wrapped with aluminum. In some cases, the aluminum may be covering some previously damaged wood. There is no thorough method to determine the extent of damage in this case. It is imperative that all caulking be maintained in these areas, otherwise water could be trapped.

Any exterior raised panels, plywood panels, or garage door panels must have the caulk integrity maintained, or moisture will infiltrate and cause deterioration. Garage door jambs are highly prone to rot due to ground or driveway contact. The wicking of the water through capillary action is a common source of moisture. (See Fig. 4E) This is also a potential area for termites to migrate. Either sealing the end grain pier or cutting the bottom of the jamb up to 3/8" will break the wicking action.

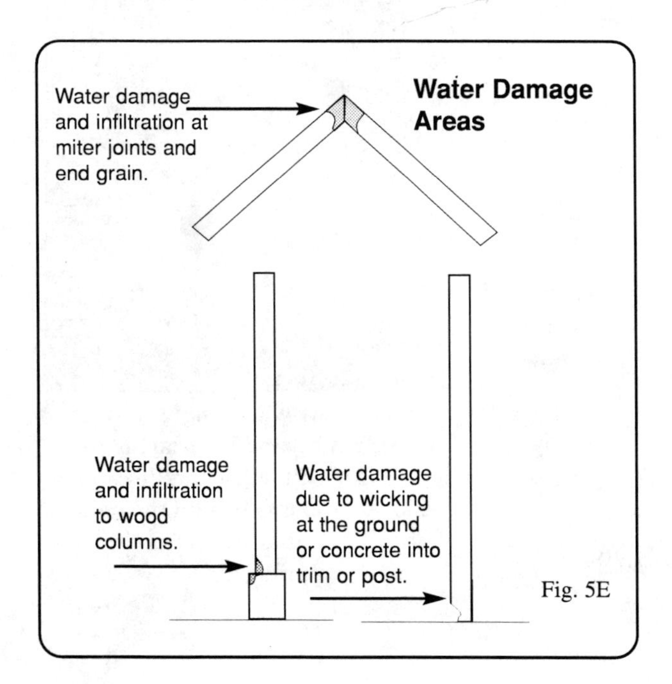

Caulking is probably more important to maintain at the end grain or butt joints of the trim. If moisture infiltrates these points, the paint may prematurely fail (this is an indicator of moisture) or the moisture may not freely escape and water damage or rot will occur. If any signs of fungus are present, that could be another indicator of moisture in the wood. (See moisture problems under the "Structure" chapter). If the wood had the end grains sealed prior to installation, the likelihood of water infiltration or rot drastically decreases.

 ***Maintenance may include:***

- Caulking at all connections to windows, doors, other trims, brick, wall penetrations, end grain, butt joints, etc.
- Maintaining all other sealants or paints.
- Keeping the grading from contact.

Railing

Wrought iron railing is highly susceptible to rusting, especially at the base. If the stem is set in concrete the possibility of rusting is drastically increased. By caulking around the base and creating a mound, the moisture can be diverted away and may help prolong the life of the railing. If the railing has been rusted off, a new base may be welded on some railings.

Wooden railing is susceptible to all the usual wood rot problems.

Basement Stairs

With a stairwell there should be a drain at the bottom of any exterior stairs. If there is not, water could back up into the basement. This drain must be kept clean of any debris. The drain on an older home might simply be a french drain (gravel pit), but newer homes many times have a drain into the interior sump pump. Problem cases might rely on a covering to prevent the water from entering the stairwell. Usually a block wall acts as a retaining wall as well, and many times there are no weep holes to allow water to drain. In some cases this water pressure could cause the wall to crack and eventually fail. Weep holes may help prevent this occurrence.

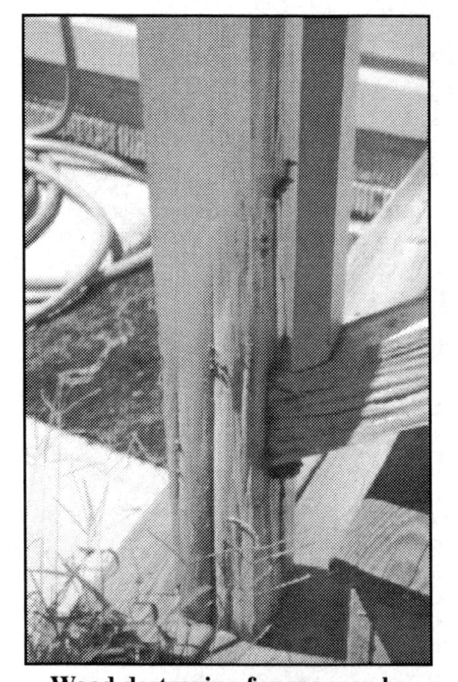

Wood destroying fungus may be evident by wood appearing to collapse within itself. mildew or black speckling behind the paint may be an indicator of moisture infiltration.

⚠ Power Wash Warning

High pressure from power washers may cause severe damage to siding, decks, roofing material and other surfaces. Enlarged cracks, knots may come loose and raised grain. Water infiltration and permanent staining may result.. Masonry may also be adversely affected by power washing. Low pressure from a garden sprayer or similar and bleach or other cleaner diluted will typically clean mildews, algeas and other growths.

EXTERIOR CHECKLIST

1] **SIDING MATERIAL** ☐ ***At or Below Grade, Susceptible to Rot, Insect, Water Infiltration & Damage.***[1] ☐ **Exposed Wall Framing, Sheathing**[1]
☐ **Siding Rows Sagging, Possible Settlement**[1] ☐ **Undulation Noted Sighting Along Wall**[1] ☐ **Bulge In Wall (*Monitor for activity*)**[1]

a] ☐ **VINYL** ☐ **Problem;** Loose / Cracked / Buckled / Missing Section(s) / Amateur, Sloppy Installation / Incomplete in Areas / Holes

b] ☐ **ALUMINUM** ☐ **Problem;** Loose / Dented / Chalking / Missing Section(s) / Amateur, Poor Installation / Previous Repairs / Incomplete in Areas

c] ☐ **WOOD** **Style;** Clapboard / Dutch-lap / Beaded / Wood Shingles, Shakes / Grooved Plywood / Board & Batten / T & G / Vertical / Diagonal
Material ☐ **Cedar, Redwood, Cypress** ☐ **Pine, Fir** ☐ **Composite, Hardboard** *(May be susceptible to delamination and failure)* ☐ Unknown
☐ **Problem;** Rot / Water, Insect Damage / Curling / Delaminating / Cracking / Loose / Missing Sections / Ground Contact[1] / Knots Loose, Missing Inadequate Installation / Improper Nailing Backing / No Visible Air Barrier / Nails Bleeding / Double Nailing / Inadequately Secured / Deteriorated

d] ☐ **BRICK / STONE** ☐ **Problem;** Spalling / Efflorescence / Water Damage / Loose Section(s) / Cracked Section(s) / Amateur, Sloppy Installation Veneer Separating From Structure / Inadequate Sill Pitch / [**Mortar Joints;** Failing, Cracking, Amateur Point Up, Previous Repairs, Deteriorating]

e] ☐ **STUCCO** ☐ **Problem;** Cracking / Crumbling / Failing / Separating from Wall / Sealant at Wall Penetrations Failing / Moisture Infiltration

f] ☐ **SHINGLES** ☐ **Cement Board** *(Possible Asbestos)* ☐ **Fiberglass** ☐ **Composite, Hardboard** ☐ **Asphalt** ☐ **Metal**
☐ **Problem;** Cracked / Missing / Failing / Damaged / Deteriorating / Rusting / Loose Section(s) / Delaminating / Amateur, Sloppy Installation

g] ☐ **FORM STONE** ☐ **Problem;** Cracking / Crumbling / Failing / Separating from Wall / Sealant at Wall Penetrations Failing / Moisture Infiltration

2] **PAINTS & SEALANT(S);** ☐ N/A ☐ **Clear** ☐ **Stain** ☐ **Paint** ☐ **Problem;** None / Failing *(Susceptible To Water Infiltration and/or Damage)* / Weathered

3] **CAULKING**[2] ☐ Ok ☐ **Problem;** Missing / Cracking / Failing / Signs of Water Infiltration ☐ ***Updating Recommended To Prevent Water Damage***
☐ **Problem;** ***Caulk Needed At;*** Windows / Doors / Trim / Siding Intersections / Electrical Wires / Heat Line / Plumbing / Light Fixture(s) / Dryer Vent

4] **TRIM**[2] ☐ Ok ☐ **Wood**[2] ☐ **Vinyl** ☐ **Aluminum** ☐ **Vinyl Clad** ☐ **Aluminum Clad** ☐ **Stucco** ☐ **Hardboard, Composite**[2]
☐ **Problem;** Rot / Loose / Missing / Ground Contact / Sealant Failing[2] / Susceptible To Water Infiltration and/or Damage / Amateur, Sloppy Installation
☐ **Problem Located;** Fascia / Rake / Drip Cap / **[Window(s);** Sill, Casing, Frame], / **[Door(s)**; Toe Kick, Jamb / Casing / Head Trim / Plinth Blocks]

5] **SOFFIT** ☐ N/A ☐ **Wood** ☐ **Vinyl, Aluminum** ☐ **Exposed Rafter Tails** **[VENTED;** ☐ No ☐ Yes ☐ Continuous ☐ Partial]
☐ **Problem;** Rot / Water Stains / Falling Down / Missing Section(s) /Amateur Installation / Inadequate Venting*(Upgrade)* / Missing, Damaged Screens

6] **COLUMN(S)** ☐ Yes ☐ **No** ☐ **Wood**[2] ☐ **Metal** ☐ **Problem;** Rot / Loose / Ground Contact / Sealant Failing[2] / Rusted / Failing

7] **GUTTERS** ☐ Yes ☐ **No** ☐ **Aluminum** ☐ **Plastic, Vinyl** ☐ **Copper** ☐ **Galvanized** ☐ **Wood** ☐ **Scuppers**
☐ **Problem; [*Missing*;** Gutter Section(s) / Downspout(s) / Splash Guard(s) / Drain-tile / Elbow(s) / Screens*(Recommended)* / Support Brackets, Spikes]
☐ ***Need Cleaning*** ☐ **Inadequate;** Installation / Pitch For Drainage / Angle to Roof / Size Of Gutter / # of Downspouts / Location Of Downspouts

8] **DRIP EDGE** ☐ Yes ☐ No ☐ **N/V** ☐ ***Recommended*** ☐ **Problem;** *Rot at*; Sheathing, Fascia / *Susceptible to*; Water, Ice Damage

9] **ELECTRICAL** ☐ Yes ☐ **Problem; Receptacle(s);** Not Operational / Not Located / Recommended GFCI / Hazardous / Loose, Damaged Cover

10] **HOSE BIBS** ☐ Yes ☐ **Problem;** Turned Off or Not Operational / Not Located / Damaged / Handle Missing / At Grade *(Possible Backventing)*

•**ADDITIONAL COMMENTS** *[1]Any signs of rot, water infiltration, insect damage, settlement, or abnormalities should be repaired by an appropriate licensed or qualified individual or company. [2]The siding should have several inches of clearance from the ground, or water and/or insect damage may occur. [3]All exterior caulks, paints and sealants need to be maintained to prevent water infiltration and potential damage. [4]Ground Fault Circuit Interrupters (GFCI) are recommended at outside receptacles.*

____

____

____

____

____

____

____

____

____

____

☐ ***Visible Rot, Water, Insect Damage***[1] ☐ ***Further Evaluation & Repairs Required***

Decks & Porches

Your deck is an area with several recurring problems. A deck is subjected to extreme weather conditions, i.e. rain, snow, heat, sun, etc., and requires routine maintenance and inspection. Large numbers of people may congregate on a deck, and it is consequently subjected to large weight loads. An interior floor will usually have direct bearing points for the floor joist, whereas a deck may be supported only by connectors at the house and/or support post.

Quite commonly the ledger board (structural connection supporting floor joists connected to the house) is attached directly to the house, with no flashing or spacing to prevent water infiltration. This can lead to rot at the sill plate, bandboard, ledger, and/or siding. Water can also pocket and freeze, causing damage to the siding --- even vinyl and aluminum. (See Fig. 1D).

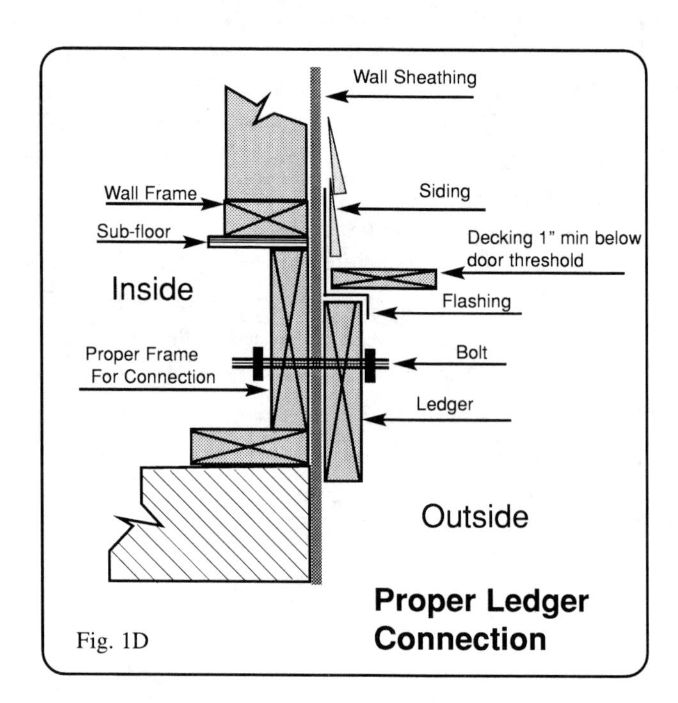

Fig. 1D **Proper Ledger Connection**

There are some recommended methods for preventing water infiltration. One solution is a flashing cap. The flashing should go up under the siding at least several inches, angle out on top of the ledger, and bend down over the face of the ledger. This will force the water to run off and away from the siding. Another acceptable method is to space the ledger from the house, and fill the bolt holes with silicone or some other sealant prior to inserting the bolts. This will allow the water to drain between the deck and siding. You could also build a free standing deck with no attachments to the house.

Erosion can undermine footings.

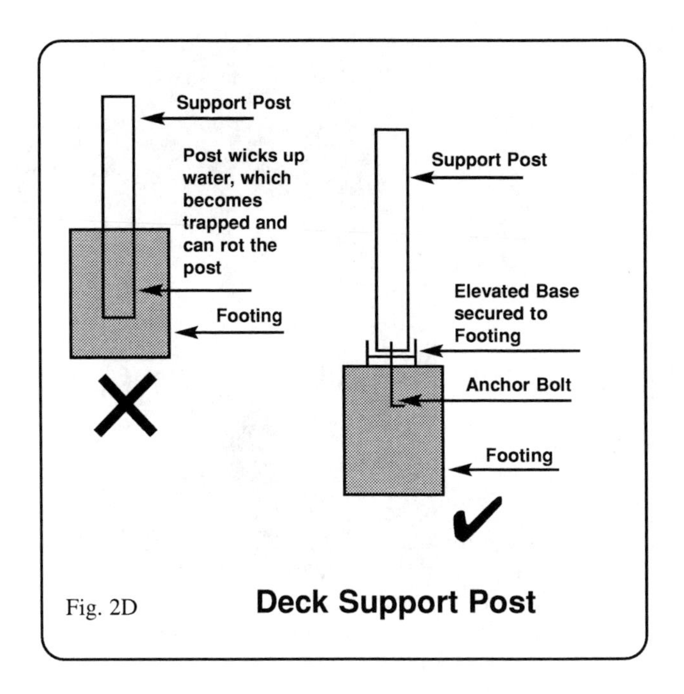

Fig. 2D

Some houses are constructed with pre-engineered floor joist. Many times these floor systems will have a plywood or other non-dimensional (pre-engineered) framing band-board. In this instance, a deck attached to the plywood is susceptible to failure and potential collapse, as has been documented in many occurrences. Some areas are requiring that a deck be independent of the house structure, and rest on independent supports and footings.

Another serious problem occurs when support post are put into the footing, or even on the ground, where they are prone to rot. (See Fig. 2D) Some people hold the misconception that pressure treated wood will not rot, but it may completely fail within a few years, as will 6x6's. When a post is put into the ground and concrete is poured around the post, the post has established a pocket in which water can collect. The water is absorbed into the post and cannot dry out. In colder climates the water freezes and can crack the concrete. Additionally, the deck should be bearing on top of the footing and not in it.

Support post should be centered on footing.

One preferred method for defeating this problem is to pour the footing either level or above the grade. Then, using a positive connection, such as a galvanized anchor base, the post should be bolted or embedded in the concrete. This will elevate the post off of the ground, and out of the footing, and will prevent the wood's capillaries from wicking up water. In addition, you will have direct bearing on the footing. It is also acceptable to pour a footing below the frost line, at least 8" thick, then sit the post on top of the concrete and tamp fill dirt around the post. This will allow the water to drain away from the post. Some soils may still cause the post to rot, so the post needs to be routinely monitored. If your

Top of stair stringers inadequate connection susceptible to failure.

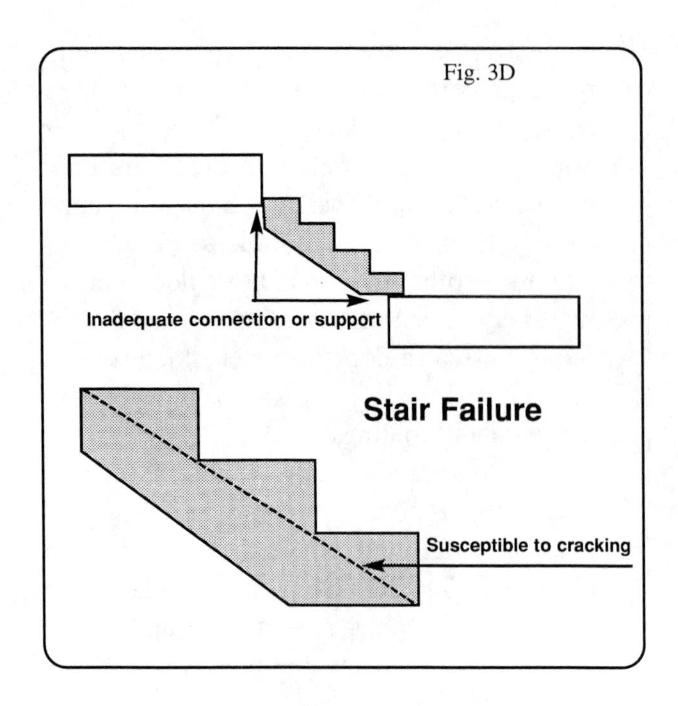

Stair stringers separating from treads are susceptible to failure.

posts are in poured concrete it is recommended to routinely inspect the post with a sharp probe for any rot.

Other items to visually check for are a positive connection at the top and bottom of the stairs to the deck or landing. Quite commonly the connections are inadequate. With notched stringers, it is imperative to have the bottom section of the stringer, which is solid, connected to the deck. If you draw a parallel line with the top edge of the board, the connection should fall in the solid section of the board. Otherwise, this connection is prone to failure if the grain should crack or fail. (See Fig. 3D).

> *Definitions...*
>
> **Triangulation:** Structural application using triangular shapes distributing compression, tension and loads used in floor and roof systems.

See if all of the railings and pickets are secure. We prefer to see the railing post through bolted. All railing should be bolted to the deck frame securely. On higher decks, check for railings wide enough to allow a child to fall through. All pickets should be secure. A high deck with decking running perpendicular to the joist may be susceptible to shaking back and fourth. This may be due the lack of proper triangulation. A board fastened to the bottom of the joist from the ledger to the outside bandboard can help stabilize the deck, and decking running on a diagonal can add a tremendous amount of triangulation to the deck.

Diagonal decking is attractive and adds much needed triangulation to prevent swaying.

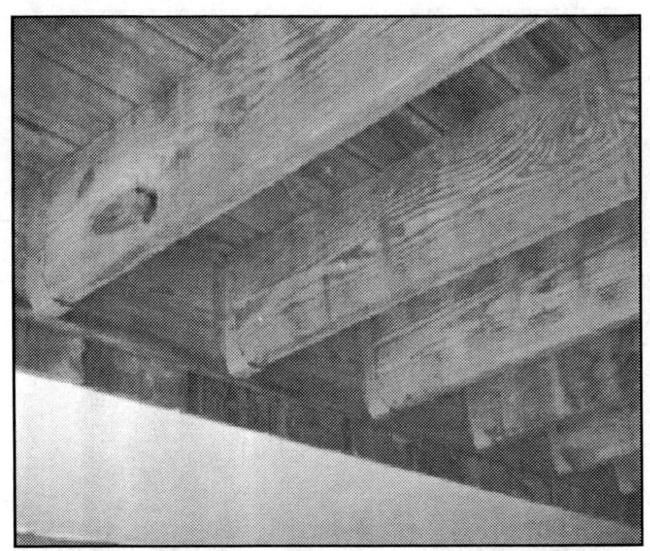

Decks should be bolted to house with properly installed joist hangers and a flashing detail to divert water away from the structure. (Flashing missing above.)

MAINTAINING YOUR DECK

There are a few schools of thought on maintaining a deck. One is to let the deck weather naturally and allow the wood to breathe. If the wood becomes discolored or has a mildew problem, you can use a mild mix of bleach and water to kill the mildew. There are chemicals on the market you may purchase, but some of them are harmful to the wood, and can cause the grain to raise. Be careful with power washing your deck; the pressure can severely raise the grain of the wood and cause damage. We know of one case where, due to the water pressure, a deck swelled so much that once the decking dried out, the nails popped out one quarter of an inch.

Decks may prematurely rot if they are in conditions where they remain damp all of the time. Some people will seal the deck to prevent moisture intrusion, thus repelling the water and keeping it from entering into the grain and causing damage. Your decision should also be determined by the height of the deck. If the deck is too low to the ground, moisture may become trapped in the wood. The sealer can act as a barrier and not let the moisture escape, causing mildew and fungus to become trapped under the sealer in the wood, leading to rot.

You should determine what species of wood you have. Is it treated, cedar, redwood, cyprus, fir or a wood that may be more susceptible to rot such as spruce or white pine? You may want to consult with a local paint supply company for the best product for your type of wood.

Install missing joist hangers at ledger to ensure proper connection.

There are many sealers on the market so shop around and educate yourself. Some sealers have ultraviolet protectors which help keep the sun from fading the wood. There are translucent stains which have a pigment, which allow the grain to remain visible and still protects the wood. Stains normally last longer than the clear sealers and can bring color back to a faded deck. The clear sealers will show the wood's beauty, but depending on the deck's location, exposure to the sun, dampness and traffic will affect the sealer's life.

Deck floating on flat roof should have proper drainage and air circulation. There should also be a method to access roof for inspection and repairs.

You should routinely inspect your deck for rot, decay, cracking, settling and severely opening the grain. Check all of the connections at the rail, ledger and stairs. Examine the stairs for cracking or failing treads. If your posts go below grade, look for any rot by using an awl or a sharp probe. If you suspect any problems call a professional to come and inspect the problem.

PORCH & DECK CHECKLIST

1] **OVERALL CONDITION** ☐**Adequate** ☐**Fair** ☐**Poor** ☐**Amateur / Sloppy Workmanship** ☐***Hazardous Condition***

2] **FOOTING(S) VISIBLE** ☐Yes ☐No ☐***Support(s) Post Go Below Grade, and are Susceptible to Rot, Monitor Routinely***
☐**Post Inadequately Fastened to Footing** ☐**Inadequate Footing(s), Recommend Updating** ☐**Repairs Required**

3] **SUPPORT POST(S)** ☐**4x4 / 4x6 / 6x6** ☐**Concrete Piers** ☐**Brick Piers** ☐**Metal Columns** ☐**Other**________________
☐**Problem;** Cracked / Twisted / Rotted / Inadequate / Checking / Bowed / Rusted / Leaning / Failing / Inadequate Connection to Frame, Beam

4] **FRAME** ☐Ok ☐**Problem;** Requires Additional Support / Single Beam / Bouncy / Sagging / Shaky, *Needs Diagonal Bracing* / Rotted / Failing Separating From Structure / Inadequate / Loose, Weak Boards / Nails Rusted / Excessive Span / Excessive Spacing / Hazardous / No Access Below

a] **Joist Dimension;** ☐**2x4** ☐**2x6** ☐**2x8** ☐**2x10** ☐**2x12** ☐**Beams** ☐**Excessive Cantilever** *(Bouncy)*

b] **Material;** Treated / Cedar / Redwood / Metal / Plastic / Concrete Deck / Concrete Beams / Composite / Unknown / Painted, Stained

c] **Joist Hangers** ☐Ok ☐**Problem;** Missing / Inadequately, Improperly Nailed / Missing Corner Connections at End Joist to Ledger / Severe Rust

d] **Decking Dimension;** ☐**1x4** ☐**1x6** ☐**5/4x6** ☐**2x4** ☐**2x6** ☐**2x8** ☐**Tongue & Groove** ☐**Plywood**

e] **Material;** Treated / Cedar / Redwood / Plywood / Non-Treated Wood / Metal / Plastic / Indoor-Outdoor Carpet / Unknown / Painted, Stained

f] **Condition Of Wood** ☐Good ☐Fair ☐**Poor** ☐**Problem;** Rot / Splitting / Curling / Cupping / Checking Excessive Decking Span

5] **RAIL** ☐Yes ☐N/A ☐**Recommended** ☐**Inadequate** **Material;** Wood / Metal / Plastic / Other________________
☐**Problem;** Loose or Missing Pickets / Poor Condition / Loose / Warped / Rusting / Failing / Not Bolted / Inadequate Height / Hazardous

6] **LEDGER** **Bolted** ☐Yes ☐No ☐Required ☐N/V **Flashed** ☐Yes ☐No ☐*Recommended* ☐N/V
☐**Problem;** Visible Rot / Susceptible to Water Infiltration / Siding Damaged / Not Properly Secured ☐**Repairs Required**

7] **AIR CIRCULATION** **Under Deck;** ☐Ok ☐**Problem;** Restricted / Deck at or on Ground / None ☐***Recommend Updating***

8] **PROPER DRAINAGE** **Under Deck;** ☐Ok Dirt / Concrete / Grass / Filter Cloth / Ground Cover / Brick / Gravel / Other________________
☐**Problem;** Damp / Standing Water / Erosion / Mud / Undermined Footings ☐***Recommend;*** Grading Away From Structure / Filter Cloth / Gravel

9] **STAIRS** ☐Ok ☐**Problem;** Improper Connection at Top, Bottom / Rot / Cracked or Inadequate Stringers ☐**Repair / Replace**

10] **PATIO** ☐Brick ☐Concrete ☐Gravel ☐Tile ☐Wood ☐**Problem;** Poor Condition / Cracking / Water Trap / Settlement

11] **DECK / PATIO COVER** ☐Pitched Roof ☐Flat Roof ☐Shade Trellis ☐**Poor Condition** ☐**Repair / Replace**

12] **ROOF DECK ON SLEEPERS** ☐Ok ☐**Problem;** Not Accessible for Inspection / Rot / Failing ☐***Recommend Further Evaluation***

13] **ROOF DECK** ☐Ok ☐**Problem;** Not Accessible for Inspection / Rot / Failing / Inadequate / Leaks ☐**Repair / Replace**

14] **HOT TUB / SPA SUPPORTS** ☐Ok ☐**Problem;** Inadequate / Hazardous / Recommend Further Structural Evaluation

15] **SCREEN CONDITION** ☐N/A ☐**Problem;** Torn / Deteriorated / Missing / Screen Door Damaged ☐**Repair / Replace**

•**ADDITIONAL COMMENTS** **Multiple decks are incorporated in one report. Decks are subject to extreme weather and the condition of the wood can change drastically with time. Maintenance and routine inspections of your deck and components are recommended. Amateur work should be repaired and upgraded.*

____
____
____
____
____
____
____
____
____
____
____
____

☐***Further Structural Evaluation & Repairs Required*** ☐**Updating and/or Replacing Recommended**

Roofs

> *Definitions...*
>
> **Pitch:** The pitch of the roof is determined by the rate the slope rises. For example, a 4/12 pitch means for every 12" of run (horizontal), the roof rises 4" (vertical).

When viewing a roof, you must decide whether you need to walk on the roof for a close inspection or simply view the roof from the ground with binoculars. Sometimes a combination of both choices may be necessary. Remember: you need to exercise common sense and good judgement when you decide to get *on* a roof. There might be an occasion when the roof cannot be safely accessed by one person with a ladder.

Before stepping on the roof, ascertain its material and structural integrity. If the shingles are in poor condition, walking on them could cause the shingles to crumble under your feet. Slate should never be walked on; other materials such as metal are inherently slippery. If the roof is wet or icy it should be viewed from the ladder. Whether or not you get on a particular pitched roof is a personal decision, based on your own skill and agility. Keep in mind that conditions on a roof may change the accessibility of various pitches. For example, on a hot day, asphalt shingles on a 6/12 pitch can become slick.

 ***Reasons <u>not</u> to get on a roof:***

- Your own skills and abilities may limit you.
- Is the ladder stable? Can it slip out from under you? Is someone holding it?
- Is the roof too high to safely access?
- The roof may become slippery when wet, or when covered with moss, ice, etc. The material could be inherently slippery.
- Is the roof hazardous due to weak structure and could it collapse? Could the roofing material break off under your feet and cause you to slide off the roof?
- Walking on a roof such as slate could damage the roofing material.

 ***Reasons <u>for</u> getting on a roof:***

- The inability to see potentially hidden problems from the ground.

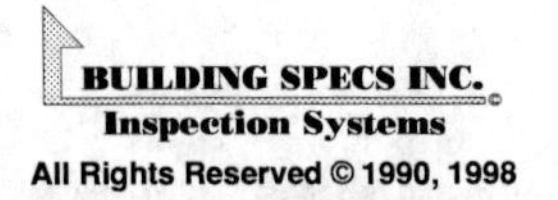

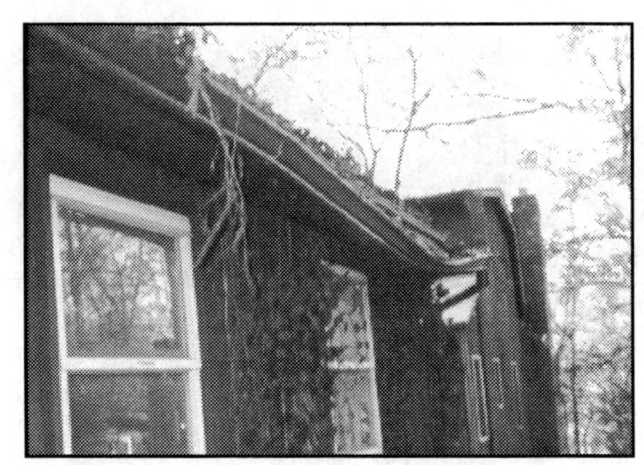

Inspect gutters for excessive debris and proper connections to fascia.

- To inspect flat roofs.
- To inspect components such as plumbing vents, skylights, etc., up close.
- To inspect the roofing material up close.
- To walk the roof in order to feel for weak or soft spots and structural integrity.
- To inspect chimney caps.

If you get on the roof, look at the sheathing if it is visible for any rot or water damage. Also check the condition of the gutters to see if they are full. When gutters are filled with leaves, the leaves act like sponges retaining water, allowing it to soak into the sheathing (solid or spaced wood fastened to the top of the roof frame for roofing material).

When inspecting the structure, look for rot or severe sagging of the ridge or rafters. This could be an indication of an unsafe roof. Lightly bounce once on the roof; this may show some of the structural integrity of the frame. You can also view the framing from inside if the attic is accessible.

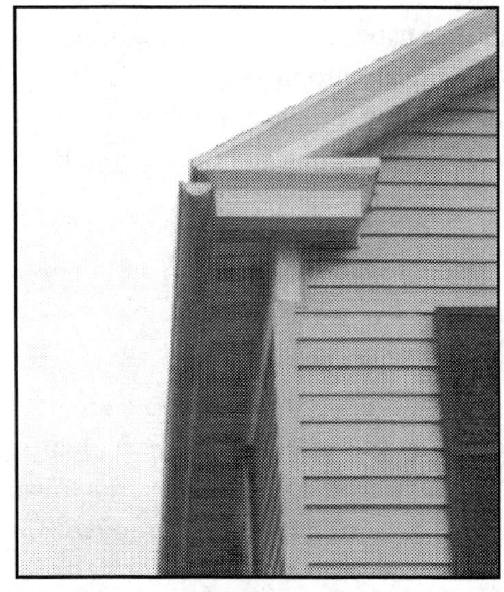

Gutters installed at improper location for drainage allows water to overshoot, spilling directly to the ground.

Dormers and gambrel roofs are prone to leaking and water damage.

 ***Roof Irregularities:***

- *Sagging Ridge*: Missing or inadequate collar ties, lack of proper triangulation ties, poor workmanship.
- *Sagging rafters*: Excessive span, cracked rafter, reversed crown during installation.
- *Hump*: Excessively crowned rafter, factory truss imperfection, poor framing alignment.
- *Bouncy Areas*: Excessive spans.
- *Spongy Areas*: Possible rot, delamination.
- *Excessively weak sheathing* (possible FRT failure -- see FRT section under "Attics").

WATER FLOW

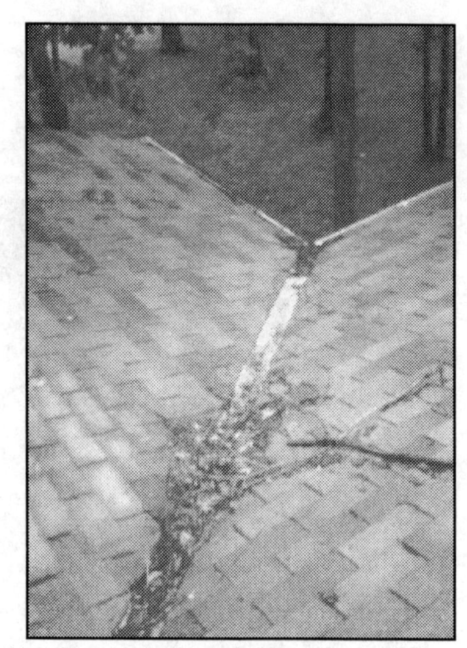

Keep debris clear from valleys and other roof components.

When inspecting the roof you need to think about how and where the water will run. Water should always be unobstructed and flow over and on top of the lower sections as it goes. Extreme deluge as well as normal flow could allow water to backup, pond and seep into the house. The wind can also play a factor, allowing water to back up. Any areas susceptible to ice damming need close inspection. The probabilities of roof penetrations and intersections failing or leaking are greater than the main body of the roof failing. Improper installation of any roofing materials is as much a factor of premature failure as its age.

ROOFING STYLES

Roof styles may be dictated by region, climate or design. Steeply pitched roofs are widely used in northern areas to help prevent snow and ice from accumulating on top. An oceanfront house with a low pitch may be susceptible to water blowing under the shingles. Hot southern regions often utilize flat roofs due to lack of snow and low maintenance cost.

Some common roofing styles are:

- Gable
- Gambrel
- Flat
- Turret
- A-Frame
- Hip
- Mansard
- Shed
- Butterfly
- Cut-up (Valleys / Hips)

ROOFING MATERIALS

There are many different materials used for roofing, and new products are coming out every year. Materials may vary in different climates, regions and for different applications of roof styles. Hot climates utilize materials that reflect heat and do not disintegrate from the sun, i.e. metal, tile, and lighter shingles. In windy locations, pay close attention to areas and materials where water can back up and cause a leak. The material should be designed to hold up against lifting and tearing away of the roof.

Shingles may be prone to blow off when improperly installed.

A roofing material will have an estimated life warranty. This can vary due to the

following conditions:

- Weather.
- Which side of the house faces the sun.
- Color of the roofing material.
- Amount of ventilation.
- Proper installation of the roofing product.

Tree branches should be trimmed back to prevent damage.

Asphalt / Fiberglass

One of the more common materials used is asphalt shingles. Older shingles were heavy, and when they failed they crumbled. Modern shingles are lighter because they are reinforced with a fiberglass matrix which strengthens the shingle. These shingles are much thinner and may be more susceptible to surface damage.

Sloppy or amateur workmanship may allow water infiltration.

Asphalt / fiberglass shingles come in different styles and with various lifetime warranties, ranging from 15 to 40 years. This may change depending on location, ventilation and color. Darker shingles may require more ventilation to prevent heat buildup which causes shingles to fail early. Poor attic ventilation can lead to shingles literally baking from excessive heat, resulting in curling and premature failure.

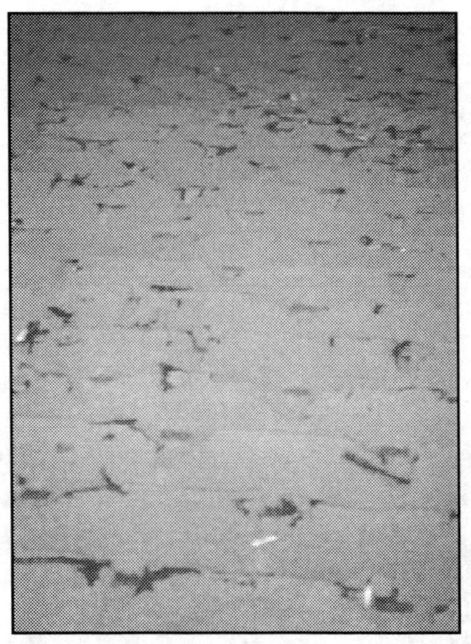

Severe erosion and curling are indicators roof is at or near the end of its life.

Three tab shingles can be identified by the way key-way lines (slots) run vertically up the roof. The key-ways are the areas where leaks are more likely to occur. The shingle below can actually become eroded through, and expose the top of the lower shingle, or expose the felt paper. Strip shingles have no key-ways and are likely to outlast 3-tab shingles. Architectural shingles are the most durable with a double overlay face and heavier weight. These shingles are meant to create an enhanced aesthetic look similar to cedar shingles and be more resistant to wind. Interlocking shingles are seen on older homes and were used before tar strips were used as a self sealant. Tar strips were implemented to help prevent wind damage.

There is very little maintenance with asphalt shingles. Keep trees trimmed back to help prevent rubbing or debris from falling on the roof. If excessive debris do accumulate, remove them. Painting the shingles is not recommended. The use of any petroleum products is not recommended on the shingles; it can cause shingles to virtually disintegrate.

Signs to look for with asphalt/fiberglass shingles:

 ***Normal wear signs :***

- Minor eroding, evident by the granular surface thinning out. Excessive granules may be found in the gutters.

 ***Severe wear signs :***

- Severe eroding.
- Fissures in the surfaces (alligatoring).
- Cracking through and across the tabs.
- Holes through the shingles.
- Curling of edges of the shingles.
- Missing shingles.
- The key-ways wearing through the shingle beneath.
- Exposed shingle matrix worn through surface, i.e., fiberglass, felt paper.

Look for poor workmanship by sighting down and up the key-ways and across the rows for straight running lines. If the rows are run too irregularly the previous shingle can become exposed at the key-way and is then susceptible to leakage.

Try to lift some tabs delicately, however, if they will not lift easily do not persist or you may damage the shingles. If you are successful in lifting the tabs, see if the shingles are properly fastened. With any of the fasteners make sure at least four fasteners per shingle are used. The fasteners should about 1" from the ends, equally divided and nailed just above the tar lines.

There are 3 common fastening techniques:

- Hand nailing with roofing nails.
- Pneumatic roofing nail gun. This type of nail gun utilizes coiled strips of roofing nails.
- Roofing staple gun. The staples this gun uses are wide, thin wire staples.

In higher wind areas, nails tend to hold better. Roofing nails are preferable to staples. If staples are used, be sure the staples are parallel with the bottom edge of the shingle. This will decrease the chance of shingles blowing off.

If there is only one layer of shingles and the roof is failing, replacing the roof may be necessary. Some areas will allow two layers of shingles to be installed; check with your local building codes. More layers can add a considerable amount of weight to the roof structure. The average square (100' sq) of shingles weighs approximately 215-400 lbs.

Three tab shingles are not recommended for roofs under a 4/12 pitch. Any lower pitch may

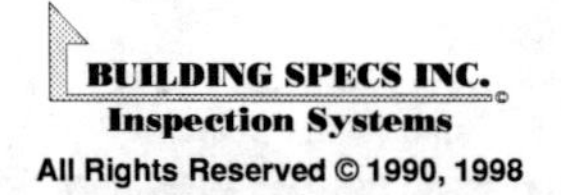

cause water to back up under the shingles especially with the aid of a strong wind. Some shingle manufacturers allow down to 3-1/2 pitch and require a 4" exposure instead of 5" and 30lb felt instead of 15lb felt.

When inspecting the roof, lift the bottom course of shingles and try to expose the sheathing underneath to look for any signs of rot, ice damage or delamination. A drip edge or ice damming membrane is highly recommended to help prevent this sort of damage in colder climates. Inspect the caps on the ridge and/or peak for cracking or poor installation. They should have close to the same exposure as the main shingles. There should not be any exposed nails, except for the last tab. Any exposed fasteners on the roof should be sealed. Look for proper shingle overhang at the gutters and rakes. The overhang should be approximately 3/4" to 1-1/4". If the overhang is too much, the shingles will sag, and over a period of time, crack off. However, if the shingles are cut too short, water could work its way under the shingles and damage the sheathing or cause a leak.

Sometimes asphalt shingles are installed directly over cedar shingles. This added weight may be too much for the structure, and can also be very aesthetically unappealing. Shingles should be installed over a flat surface, as recommended by the manufacturers. Extreme irregularities in the roof surface may also lead to premature failure in the shingles themselves.

Moss on wood shingles may lead to rot and decomposing. Roof should be cleaned and treated.

Cedar Shingles

Wooden shingles are primarily cedar, however, there are redwood, oak and a few other types manufactured. Cedar comes in two basic styles: machined shingles and split shakes. These roofs can be very slippery and susceptible to loose shingles, so exercise good judgement.

Machined shingles are thin and smoothly cut. They can be applied over open slats to allow for air circulation and drying. They may also be installed with felt paper over a solid sheathing. Some techniques include plywood sheathing overlaid with felt paper (tar impregnated) with slats laid on top. The shingles are then installed so they will have air circulation below. The life expectancy for machine shingles can be 30 to 40 years with proper ventilation and care.

Thicker split shingles are applied over a layer of felt paper, with half layers over a section of every row. They may also be applied on plywood sheathing overlaid with felt paper with slats laid on top. Hand split shingles' life expectancy is normally 10 to 15 years. If there is a moisture or ventilation problem, their life could be shortened.

As with asphalt, wooden shingles require a minimum pitch of 4/12 to prevent water penetration from backing under. The manufacturer will require a specific exposure for different size shingles, which may vary from 5" to 10".

Some wooden shingles are treated with a fire resistant chemical. You may want to maintain the shingles with sealers, or let them weather grey. The caps should have a metal or felt strip under them to prevent water infiltration.

 ***Problems to look for: CEDAR SHINGLES***

- Excessive splitting
- Poor workmanship
- Excessive nails exposed
- Rot
- Curling
- Excessive moss
- Open knot holes
- Holes through the shingles
- Missing or loose shingles

High wind areas might require a drip edge or sealer along the rakes (sloping trim section adjacent to the roof line). There can be a gap at this area, especially with splits, and water can blow in this area and result in a leak. Sight the shingles across to look for straight rows or sloppy applications. The shingles should be nailed with 2" long ring shank nails, snug, and not too tight. This allows for proper expansion and contraction. Some are stapled tightly with pneumatic staples; this could lead to premature splitting of the shingle.

Maintenance should include repairing the roof as needed by replacing any damaged or split shingles, removing debris, removing any moss, and repairing any rot damage. If moss is present, there are fungus inhibiters available to help deter growth, and repeated treatments may be required. There are also preservatives available for wooden roofs. If any shingles are severely curling or lifting, they may need to be nailed down and the exposed nails should be sealed. Look for adequate overhang of approximately 1-2" for the gutters and rakes.

Slate Roofing

Slate roofs can have a life span of 50 years or more under proper conditions. Since they are very heavy shingles, the structure has to be designed to accommodate this weight. These shingles should not be walked on if at all possible. They are susceptible to breaking, and could be hazardous to the person who has a shingle break underfoot. You might compare it to downhill skiing, but faster. Also, the damage incurred to the roof can be very costly.

Slate roofs should never be walked on, or damage may occur.

Inspection from a ladder or on the ground with binoculars is recommended. You could also rent a lift platform which will get you closer to the roof if you feel that is necessary.

Slate should be installed on slightly steeper pitched roofs, preferably an 8/12. Many times the flashings will fail before the roof will, so pay close attention to valleys and other flashing areas. The ridge may consist of only the tops of the shingles butted together and sealed with caulk or tar. Some will have a metal cap running along the ridge. These are usually routinely painted. Another way to seal the peak is to install slate caps along the ridge or hips.

 ***Problems to look for: SLATE***

- Broken or damaged shingles. Inspect the edges of the shingle for excessive deterioration or wear. Look for surface cracking or the layer of the slate actually failing.
- This roof should never have another layer applied over it; the first layer of slate should be removed before installing a new roof.
- Check to see if ice sheeting clips are installed. These prevent sheets of ice from sliding off and possibly injuring someone.

 ***Maintenance may include:***

- Keeping debris off roof.
- Replacing any cracked or broken shingles.

Tile Roofing

Tile roofs are another very durable, long lasting type of roof, which normally lasts 50 years or longer. Tile roofs are usually made from terra cotta or cement. They are also a very heavy material, so the structure needs to be designed accordingly. Maintain by keeping the roof free from debris, replacing any loose or damaged tiles, and repairing any loose cement. These roofs can be very slippery, so, again, exercise good judgement.

 ***Problems to look for: TILE***

- Inadequate structural support.
- Broken or missing tiles.
- Amateur installation.

 ***Maintenance may include:***

- Keeping debris off roof.
- Replacing any cracked or broken tiles.

Metal Roofing

Metal is becoming more popular, especially with the increasing number of styles and colors available. It is also a very durable product requiring little maintenance; the life expectancy for metal roofs can be 20-40 years, and in some cases, more. Their weight load on the structure is nominal, however, special framing techniques will need to be implemented. This usually requires cross slats perpendicular to the rafters. These roofs can be extremely slippery, so exercise good judgement.

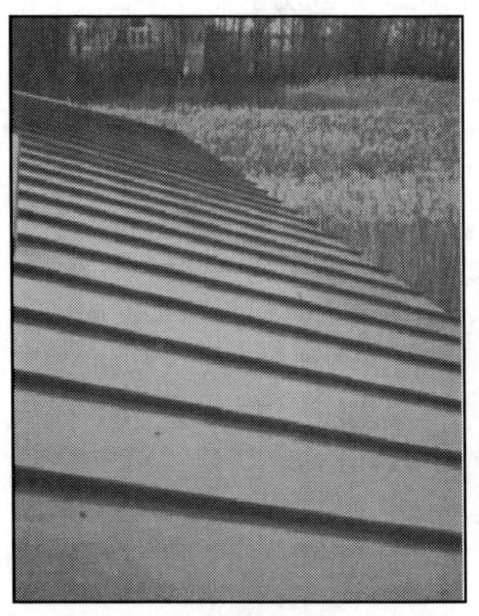

Anodized metal standing seam roofs are virtually maintenance free.

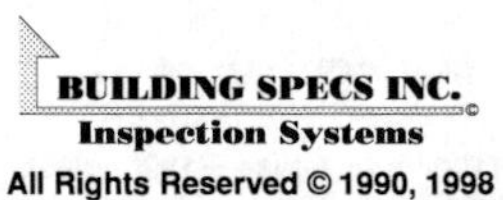

There are several styles of metal roofs being fabricated:

- Standing seam
- Ribbed sheets
- Corrugated
- Stamped (Shingles)

Any of these may be anodized, galvanized, or painted. The materials used are aluminum, copper, tin and steel.

 ***Problems to look for: METAL***

- Rust or corrosion.
- Pitting.
- Sections lifting (loose).
- Fasteners backing out.
- Improper installation.
- Installed on a roof with too low of a pitch.
- Holes through panels.
- Finish in poor condition (Paint, tar coating, anodized coating).

 ***Maintenance may include:***

- Keeping debris off.
- Removing rust or corrosion.
- Maintaining tar coating or other sealers.
- Filling holes and fastening loose panels.

Flat Roofing

Flat roofs are widely used on commercial buildings, but also often seen on residential homes in some areas. Inspection from on top of the roof is a good idea, since viewing from the ground is limited at best, but access is limited to some flat roofs due to their excessive height. The life expectancy of these roofs can vary from 10-20 years.

There are a several different materials used on flat roofs:

- Rubber membrane (Bitumen membrane)
- Roll Roofing (asphalt, mineral surface)
- Roll Roofing (asphalt, smooth surface)
- PVC Fabric
- Built up Asphalt (smooth)
- Built up Asphalt (gravel)
- Metal

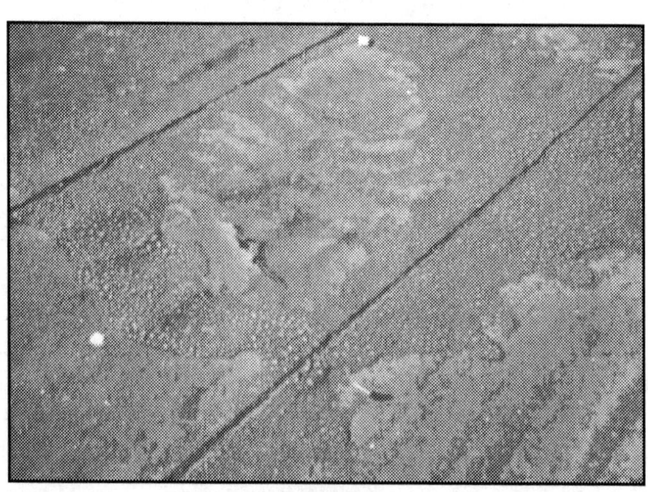

An improperly maintained flat roof may lead to failure of the roofing fabric.

Rubber Membrane (modified bitumen)

This can be a very durable product and has great application flexibility. It can be used to cover large areas and tends to stay bonded at the seams. The edges should be sealed on top of a drip edge (termination strip). Holes can be easily repaired. Potential trouble areas can be easily fortified by building up several layers. The membrane can easily be used as a flashing, incorporated into the main section if needed.

 ***Problems to look for: FLAT***

- Seam separation.
- Dry rot of the fabric.
- Sloppy installation.

 ***Maintenance may include:***

- Keeping debris off roof.
- Keeping edges, seams and roof penetrations sealed.

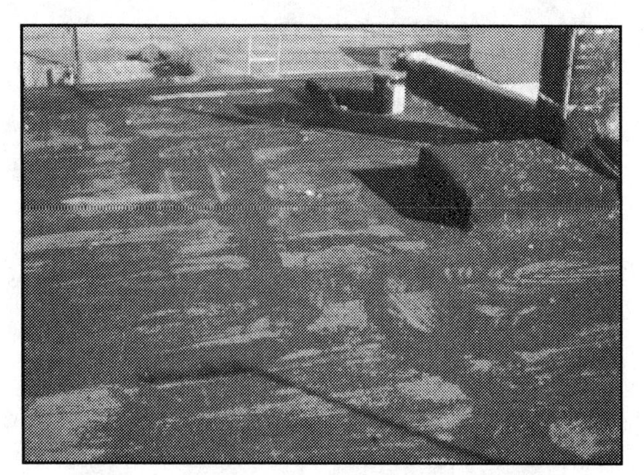

Evidence of a properly maintained flat roof.

Roll Roofing

This material is similar to an asphalt shingle. It may come with or without the mineral surface. The runs should be run sequentially starting from the bottom and moving upward. The roll roofing should overlap a minimum of two inches, and in some instances a third or half overlap may be required. Roll roofing should be used on roofs that have some pitch, approximately 1/12, but may be seen on completely flat roofs with proper installation procedures (more of an overlap). On any pitch less than 3/12 the overlaps should be sealed completely between laps and nailed down (seal all exposed nails).

 ***Problems to look for: ROLL ROOFING***

- These roofs are prone to buckling and seam separation.
- High probability of amateur work.
- Refer to asphalt shingle failure.

 ***Maintenance may include:***

- Resealing seams, exposed fasteners, and around roof penetrations.
- Keeping debris off roof.
- Coating non-mineral roofing with liquid asphalt sealer.

PVC (Poly-Vinyl-Chloride Fabric) Roofing

This material is similar to that of an an inflatable dingy. Large sections of this fabric are overlapped and glued together. It is a very tough puncture resistant material, but may have some problems.

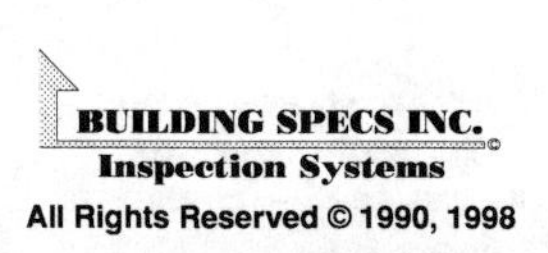

 ***Problems to look for: PVC***

- Rodents chewing holes.
- Ultraviolet light breakdown.
- Surface wearing through.

 ***Maintenance may include:***

- Resealing seams, exposed fasteners, and around roof penetrations.
- Keep debris off roof.

Built Up Roofing (With Gravel)

This material is installed in a hot liquid form over a rolled matrix and applied in several layers. A gravel layer is then applied to the surface to help protect the asphalt against the sun and any maintenance traffic.

 ***Problems to look for: BUILT UP (GRAVEL)***

- Ponding (sagging).
- Spongy areas.
- Surface wearing through.

 ***Maintenance may include:***

- Keep debris off roof.
- Replacing gravel.

Built Up Roofing (Smooth)

This material is also installed in a hot liquid form over a rolled matrix and applied in several layers.

 ***Problems to look for: BUILT UP (SMOOTH)***

- Ponding (sagging).
- Spongy areas.
- Surface wearing through.
- Alligatoring (cracking).

 ***Maintenance may include:***

- Keep debris off roof.
- Resealing any cracking with asphalt sealer.

Metal (See "Metal" roofs section.)

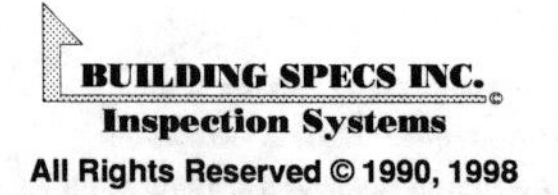

 ***When inspecting flat roofs in general look for:***

- Ponding (Indicates a low spot).
- Alligatoring (Asphalt cracking).
- Seams buckling or opening up.
- Excessive loose gravel (Built up asphalt gravel, mineral surface roll roofing).
- Rodents chewing holes (PVC).
- Holes.
- Spongy areas.
- Poor workmanship.
- Material deterioration.

 ***Maintenance of these roofs can vary and include:***

- Recoating smooth surface asphalt.
- Sealing all cracks.
- Building up low areas (ponding).
- Sealing down raised seam (seal exposed fasteners).
- Filling holes.
- Cleaning debris from roof.
- Maintaining sealer at all roof penetrations (plumbing vent, skylights, chimney, etc.), and around any drip edge.

VALLEYS

Valleys are where adjacent roof lines intersect and carry water to a point of termination. These areas should be inspected for defects, which could cause leakage.

Kinds of Valleys Include:

Open metal valley: flashing exposed, concave no bend, creased center, ribbed up at center, bent over at ends. Problems associated with this valley are holes through the metal, failing or no sealer between the shingles and metal, thin gauge inexpensive aluminum, rusting corrosion, original older flashing with newer roof, ending short of bottom of roof, and poor installation.

Closed valley: with shingle lapped or woven (with or without metal underneath). A reversed lap is when asphalt shingles are used as the valley material. One side of the valley creases into the valley and runs up onto the adjacent side of the valley approximately 12". The second side is run and cut down the center of the valley. This is a very durable valley, but installation errors are commonly seen. The roof has to be lapped so the larger or steeper roof is the second side run with the cut down the center. If it is not, the water force of the larger or steeper volume could force its way under the opposing valley. Any nails should be several inches out from the valley's center.

Woven valley: a very durable and impenetrable system, where each side of the valley has a row of shingles run and is creased into the valley and up onto the adjacent side. This alternates back

A woven valley, properly installed, is one of the most weather resistant.

and forth creating a weave. Poor installation can be this valley's main disadvantage. Look for the center of the valley weave running high onto one side of the valley. When the center rises out of the valley it may become susceptible to water infiltration.

Rubber Valley (modified bitumen): this can be a very durable product and has great application flexibility. It can be used to cover large areas and tends to stay bonded well at the seams. Holes can be easily repaired. Potential trouble areas can be easily fortified by building layers.

Keep all of the valleys clear from debris. Fill any holes and maintain the integrity of the metal valleys as needed. A sealer (roof cement) should be maintained as needed under the shingle lip to prevent water or ice from backing up under the shingles.

FLASHING

Flashings are the proverbial weak link of a roof. They are susceptible to cracks and holes that may fail before the rest of the roof does. Flashing also requires some skill and knowledge to correctly install, in order to avoid water leakage. It is common to see a new roof with the original flashing. Flashing (with the exception of copper) should be replaced at the time the new roof is installed.

Flashing is typically used to prevent water from infiltrating at two or more adjacent areas, i.e. chimneys, roof/wall connections, valleys, under the ridge and hip caps (cedar, slate), and other areas vulnerable to water, snow, or ice penetration. There are a few different styles of flashings available:

- Aluminum (different gauges)
- Copper (coated & non-coated)
- Rubber (modified Bitumen)
- Asphalt (roll and shingle)

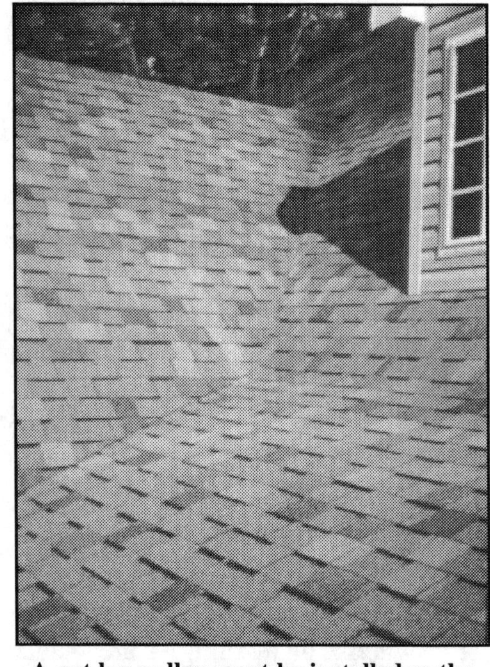

A cut lap valley must be installed so the steeper or larger roof section is placed last so water flow is diverted away.

Flashing may be utilized in many different applications, with some inherent problems for each.

Chimney Flashings

This is another area which is highly susceptible to water penetration or poor workmanship. If the roof slope ends at the back of the chimney, a cricket should be installed. The cricket is a small roof which deflects water and snow away and prevents a buildup from occurring. The cricket is usually gable shaped, with the ridge perpendicular to the chimney. The cricket can be covered with any of the aforementioned flashing materials.

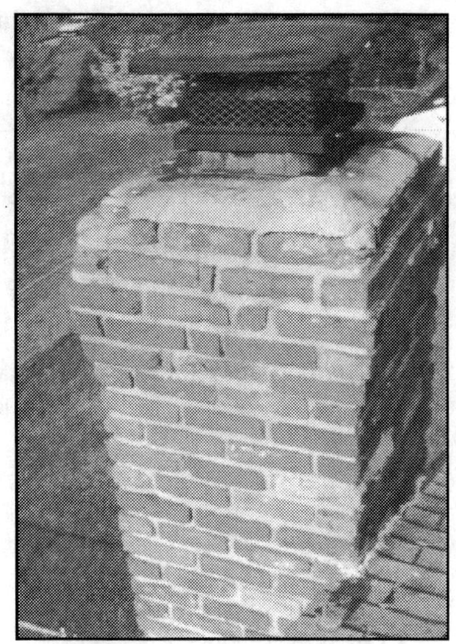

Flashing missing at chimney may allow water infiltration.

The flashing and method of installation used to envelope the perimeter of the chimney/roof connection needs to be inspected closely. The preferred method is to have the tops of the flashing turn and be sealed into a groove in the mortar joint. Some installations allow the flashing to seal tightly against the chimney. Either way, a sealant should be maintained at the top of the flashing. The lower front apron should be installed first, and the sides of the flashing should have step flashing with a counter flashing over top. The back should be the last section to be installed. It should come up at least a foot on the back of the chimney and be well sealed at the top. The ends should then neatly wrap around and over the side flashing. The section that goes under the shingles should be sealed tightly to prevent ice or water from backing up under the shingles. Trace all connections around and look for any possible place where water could infiltrate.

 ***Problems to look for: CHIMNEY FLASHING***

- Pulling away, or the sealer cracking and failing.
- Older flashings that were not updated with a newer roof.
- Weak sheathing at or around the chimney due to an on-going leak.

Routine maintenance should include maintaining the seal around the top of the flashing and removing any debris from behind the chimney.

Walls

Flashings used at wall connections are installed to create a continuous seal between two or more adjacent connections.

- *Step flashing* is installed so that each shingle utilizes an "L" shaped flashing which rests on each shingle and overlaps the lower flashing and shingle. Gravity forces the water to shed from one shingle to the next, without the aid of a sealant.
- *Apron flashing* is used at a wall which runs parallel with the shingles. The flashing is run in as long a length as possible. The flashing is tucked up behind the siding and sealed onto the top of the last row of shingles. Where the flashing meets the shingles it should be sealed, nailed down, and have the nail heads sealed.

Plumbing Vents "Boots" Flashing

These are special caps used to prevent water from getting in at the vent/roof penetration. Commonly used is a rectangular shaped cover (made from metal or plastic) with a rubber boot that fits tightly around the vent.

The bottom of the flashing should be exposed on top of the lower shingle and nailed down. This type of flashing is susceptible to the boot dry rotting, and should be inspected closely. Many times older tar repairs will be found, indicating possible problems. Covers are available that slip over the vent and protect the rubber from deterioration and from water intrusion.

Older vents may include a lead flashing. This is a lead rectangular cover with an attached tubular shape which slides over the vent. The top of the tube should be folded over into the top of the vent. These vents last a very long time, and unless a hole is evident, no maintenance is required.

SKYLIGHTS

The main purpose of a skylight is to let in sun. Unfortunately, skylights sometimes allow water to come in as well. There are several types of skylights:

- Glass or plastic.
- Factory or custom built.
- Thermopane or single pane.
- Factory flashing or on sight flashed.
- Flush mount or curb mount.
- Self flashed or step flashed.
- Tinted or clear.
- Fixed or opening.
- Manual or electronic opening.

Problems to look for: SKYLIGHTS

- Cracked or damaged pane or bubble.
- Poor installation (mounded or excessive tar repairs, especially at the top of the skylight).
- Improper skylight for application (i.e. Flat roof with skylight for pitched roof).
- Blown seal (trapped moisture, fogged, diffused).
- Functional (opening) skylight, stuck closed.
- Water stains in the well on the inside of the house evident by viewing through the skylight.

Maintenance may include:

- Keeping debris away on and above skylight.
- Maintaining any sealer around skylight.

VENTILATION

Proper ventilation can be very important to the longevity of the roof. It helps to keep the attic cooler in hot weather and allows moisture to escape. Either of these problems can contribute to a roof's premature failure. The darker the roofing material and hotter the climate, the more essential proper ventilation becomes. With asphalt shingles, the matrix and asphalt bake and dry out, leading to the shingles' premature failure, evident by curling. In colder climates, inadequate insulation in the attic can result in a problem as well. Heat rising, passing through the insulation and baking the shingles, again leads to premature failure. Also, the heat collides with cold air and forms condensation which may lead to mildew and other problems.

A turbine vent is susceptible to leaking and squeaking.

There are several kinds of ventilation available:

- Ridge (metal and plastic which requires capping).
- Power (hoods cut through, with thermostat and recommended humidistat).
- Turbine.
- Passive (hoods cut through).
- Gable (not on the roof).
- Soffit (not on the roof).

 ***Problems to look for :***

- Ridge not nailed down properly, lifting or loose (due to the nails penetrating only the sheathing and missing the framing).
- Ridge end plugs missing (seal at ends).
- Cap style ridge vent improperly nailed (too flat, needs to be at the same pitch as the roof).
- Turbine wearing out (squeaking, stuck and will not turn).
- Power (plastic hood cracked).
- Power or hood improperly installed (bottom of unit not on top of the last shingle).
- Inadequate amount of ventilation.

Very little maintenance is required other than maintaining good working condition.

ICE DAMMING

A common problem associated with winter in colder climates is ice damming. This situation can occur in several places, such as where ice is allowed to build up and where force pushes against or under another area. Gutters are probably the most common place where this occurs; snow melts from the roof and freezes in the gutter. As the ice mounds in the gutter, it may push under the shingles and onto the roof sheathing (plywood, OSB. etc.) and remain there until it melts. This can cause a leak in your ceiling or soffit overhang. If this condition is allowed to continue, rot can occur to the sheathing, frame, fascia and wooden soffit. There are preventative steps which can be taken.

Drip edge is probably the easiest and least expensive way to protect your roof sheathing. F-Style drip edge is a metal strip that comes prebent in 10-12' sections. It goes under the shingles about 2", overhangs the edge of the roof into the gutter about 3/4", and comes down over the fascia about 1", with a lip for a positive run off. This creates a seal to the edge of the roof and greatly reduces the chance of ice damming or water damage. Other methods of prevention include custom bending larger widths of flashing for greater coverage protection. The state of the art procedure is a rolled rubber type membrane that is extremely sticky on both sides and bonds to the sheathing as well as to the shingles. This has to be applied when the shingles are removed and the sheathing is exposed.

Other areas to check for ice damming are above the skylights, chimneys, valleys and adjacent walls. Preventing or repairing these areas may be as simple as sealing the vulnerable area with roofing cement. However you may want to have a professional come out to repair the roof. Some things are better left to a qualified individual.

DORMERS

Dormers are usually classified as smaller roofs protruding from the main roof, typically including a window or a vent, normally a variation of a standard roof shape. There are two main reasons for dormers: aesthetics and space. While dormers can enhance the design of the house and add to the space, there are a few areas which can cause problems

Many dormers are added to a house as an afterthought. While the majority of them are properly built, sometimes shortcuts and inadequate materials are used.

There are several dormer designs:

A Gable Dormer is a smaller version of a gable roof. It is built with the ridge perpendicular to the main roof. The ceiling line of these dormers may be either cathedral with the valley line exposed, or a flat ceiling with a buried internal valley.

 ***Problem areas may include:***

- Valley flashings.
- Apron Flashing (loose, exposed nails, missing).
- Lack of drip edge.
- Sagging at the main roof, due to load of dormer.
- Inadequate valley connection.

Flat or Shed Dormers are basically sections of the main roof lifted. The pitch and materials used may vary. Shed roofs are frequently installed as afterthoughts.

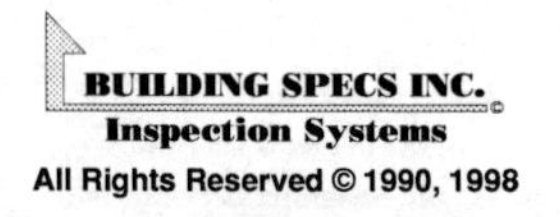

 ***Problem areas may include:***

- Sagging dormer section (inadequate framing).
- Poor connection at main roof intersection.
- Shingles on an inadequate pitch (susceptible to leak).
- Flashing along walls (step flashing).
- Inadequate or no ventilation.
- Inadequate insulation.

Hip Dormers are usually very solid and stable, and excellent for use in high wind areas. Refer to the section on gable dormer for potential problems.

Eyebrow Dormers are a beautiful focal point on a house. There is a certain amount of skill involved with installing these dormers. Framing, as well as the shingle installation, use special applications.

 ***Problem areas may include:***

- Amateur installation.
- Shingles, poor installation, susceptible to leak (low pitch).
- Valleys not blended properly.

Windows in a Gambrel Roof may have a shed dormer, a gable dormer, or be recessed with no dormer.

 ***Problem areas may include:***

- Recessed windows need special attention at the inset base at the roof penetration. This area is highly susceptible to leakage.
- The flashing around the window dormer/gambrel roof connection. This area is highly susceptible to leakage. Inadequate flashing, wood to shingle contact or extreme weather conditions may allow water to penetrate and cause damage.

FRT (FIRE RETARDANT TREATED)

FRT (Fire Retardant Treated) plywood was and is still required for a fire break at the roof for multi-family units, such as townhomes, condos, and duplexes. There was widespread failure of FRT manufactured throughout much of the 1980's and part of the 1970's. Excessive heat and moisture from improper ventilation would cause sheathing to delaminate, and in extreme cases sag in between the trusses. In some cases, leaks would develop due to this failure, and the entire roof would have to be stripped of the sheathing, re-sheathed and re-roofed.

Most areas still require a fire break, either a masonry wall that breaks the roof surface, a 5/8" fire rated drywall layer under CDX roof sheathing, or even 5/8" fire rated drywall on the ceilings of the top floor. There may be other accepted methods and variations depending on local codes. The newer FRT has been successfully updated with a buffer as well as in conjunction with updated attic ventilation. Typically, fire rated drywall or masonry is used as a firebreak between units and in the adjacent attics. When the roof sheathing breaks the plane of the adjacent roof lines a separate firebreak is required. This may be FRT or fire rated drywall sandwiched between the trusses and CDX roof sheathing.

Sometimes contractors strip the shingles and sheathing and replace it with regular CDX. This is equivalent to removing the firebreak on these units. It should be replaced with a proper firebreak.

 ***Signs of FRT failure may include:***

- Sagging.
- Delamination.
- Cracking when walked on.
- Dark discolored surface in the attic on the sheathing.
- A hairy or furry surface in the attic on the sheathing.
- Brittle, cracking sounds when you push up on the sheathing in the attic.

In cases where the sheathing is sagging, a temporary fix may may used. This involves using 2x4 blocks or larger, installed between the trusses, tight to the sheathing, in a ladder pattern. The spacing of the blocks will depend on the severity of the delamination and cracking, typically 24" on center. Another method is to rip sheets of 3/4" CDX plywood the width of the bay between the trusses. This needs to be installed tightly to the roof sheathing. The protruding shingle fasteners may impede this installation. Once the plywood is installed tight to the roof sheathing, 2x3" cleats are fastened to the sides of the truss to support the 3/4" plywood.

ROOF CHECKLIST

1] **VIEWED FROM** ☐ **Ground With Binoculars** ☐ **Walked On** ☐ **Ladder** ☐ **Limited View** ☐ **Not Visible** *(Recommend Reinspection by Roofer)*

2] **NO ACCESS** ☐ **Problem;** Too Steep / Height / Ice / Snow, Rain, Leaves, Moss / Risk Of Damage to Roof / Hazardous to Inspector / Weak Structure

3] **ROOFING MATERIAL(S)** ☐ **Multiple Layers** ☐ ***Inadequate Overhang*** ☐ ***Previous Repairs (Monitor)*** ☐ ***Not Properly Installed***

a] ☐ **ASPHALT/FIBERGLASS** ☐ Ok ☐ **3 Tab** ☐ **Architectural** ☐ **Strip** ☐ **Interlocking** ☐ Other______________

☐ **Problem;** Missing Shingles, Tabs / Eroded / Cracking / Curling / Lifting / Not Sealing / Nail Pops / Amateur, Sloppy Workmanship / Torn Exposed Matrix / Crooked Rows / Crooked Slots / Blown Off / ***Pitch Inadequate, Susceptible to Leaks*** ***(Monitor, replacement may become necessary)***

☐ **CAPS / HIPS** ☐ **Problem;** Eroded / Failing / Nearing End of Life / Curling Up / Amateur, Sloppy Workmanship / Crooked Rows

b] ☐ **WOOD** ☐ Ok ☐ **Shingles** ☐ **Split Shakes** ☐ **Problem;** *Exposed Felt Between Shingles May Lead to Premature Failure of the Shingles*

☐ **Problem;** Missing Shingles / Curling / Lifting / Splitting / Rotted / Felt Rot / Covered with Moss / Improper Installation / Crooked Rows At or Nearing End of Life / Daylight Visible in Attic at Shingles / Pitch Inadequate Susceptible to Leaks / Amateur, Sloppy Workmanship

☐ **CAPS** ☐ Ok ☐ **Problem;** Splitting / Failing / Susceptible to Leak / Crooked Rows / No Visible Water Barrier / Amateur, Sloppy Workmanship

c] ☐ **FLAT** ☐ Ok ☐ **Rubber** ☐ **Asphalt / Mineral** ☐ **Built-up Smooth** ☐ **Built-up Gravel** ☐ **PVC** ☐ **Other**___________

☐ **Problem;** Buckled / Improper Installation / At or Nearing End of Life / Ponding of Water / Inadequate Runoff / Amateur, Sloppy Workmanship

d] ☐ **METAL** ☐ Ok ☐ **Aluminum** ☐ **Anodized** ☐ **Galvanized** ☐ **Copper** ☐ **Painted, Coated** *(Unknown material, could not evaluate)*

☐ **Problem;** Rusted / Damaged / At or Nearing Near End of Life / Needs Recoating of Sealant / Loose / Amateur, Sloppy Workmanship

e] ☐ **SLATE / TILE** ☐ Ok ☐ **Problem;** Cracking / Missing Tiles / Eroded / Loose Tiles / Previous Repairs Observed / At or Nearing End of Life

4] **ROOF STYLE** ☐ **Gable** ☐ **Hip** ☐ **Shed** ☐ **Mansard** ☐ **Gambrel** ☐ **Flat** ☐ **Multi Pitch** ☐ **Other**

5] **FLASHING** ☐ **Aluminum** ☐ **Copper** ☐ **Roofing Material** ☐ **Steel** ☐ **Missing** *(Proper Installation Required)*

a] ☐ **Valley(s)** ☐ **Problem;** Reversed Lap, Improper Weave *(Susceptible to leak)* / Poor Condition / Flashing Rusted / Previous Repairs / Failing

b] ☐ **Chimney(s)** ☐ **Problem;** Improper Installation / Failing / Loose / Susceptible to Leak / Rusting / Missing / Only Caulking / Needs Caulking

c] ☐ **Step / Apron** ☐ **Problem;** Improper Installation / Failing / Loose / Susceptible to Leak / Rusting / Missing / Only Caulking / Needs Caulking

d] ☐ **Plumbing Vent** ☐ **Problem;** Improper Installation / Cracking / Loose / Susceptible to Leak / Rusting / Missing / Only Caulking / Needs Caulking

6] **SKYLIGHTS #** _____ ☐ **Glass** ☐ **Plastic** ☐ **Operable** ☐ **Self Flashed** ☐ **Amateur Fabrication,** *(Susceptible To Leaking)*

☐ **Problem;** Cracked / Fogged / Leaking / Trapped Condensation / Previous Repairs / Improper Installation / Amateur, Sloppy Workmanship

7] **VENTILATION** ☐ **N/A** ☐ **Ridge** ☐ **Power** ☐ **Turbine** ☐ **Gable** ☐ **Soffit** ☐ **Other**_________________

☐ **Problem;** Missing End Plugs / Damaged / Leaking / Secure and Seal Properly / Inadequate ***(Updating Recommended To Extend Life Of Roofing Material)***

8] **STRUCTURE** ☐ **Adequate** ☐ **Problem;** Frame Bouncy / Sheathing Weak / Visible Rot / Exposed Sheathing / Sagging / Bulge / Irregular

a] **Ridge(s), Hip(s), Valley(s)** ☐ Ok ☐ **Problem;** Sagging, *(Further Structural Evaluation Required)* / Bulge / Irregular / Hump / Weak Sheathing

9] **MISCELLANEOUS** ☐ ***Clean Debris From;*** **Gutters / Roof / Valleys** ☐ **Trim Tree Branches Back From Roof** *(Damage is or may be occurring)*

•**ADDITIONAL COMMENTS** **Severe weather conditions may cause a leak on a roof that appears adequate. The Client should have roofs which are restricted to the Inspector reinspected by a licensed roofer. Roofs and their components may require routine maintenance.*

☐ ***Visible Leaks*** ☐ ***At Or Nearing End Of Life*** ☐ ***Replacement/ Repairs Required*** ☐ ***Amateur Work*** ☐ ***Various Ages***

____

____

____

____

____

____

____

____

____

____

Attics

A large number of homes have inadequate or even no ventilation in their attics. There are no codes enforcing proper ventilation at this time, however, different manufacturers have recommendations you can follow. Depending on the climate in your location, requirements for ventilation may vary. In a four season mild climate, ventilation is important both in the summer and winter.

In the heat of summer your attic is subjected to extremely hot conditions. (See Fig. 1A) With limited ventilation such as a gable vent, only the extreme heat escapes. Many times the wood around the vent will become dark and discolored from the heat.

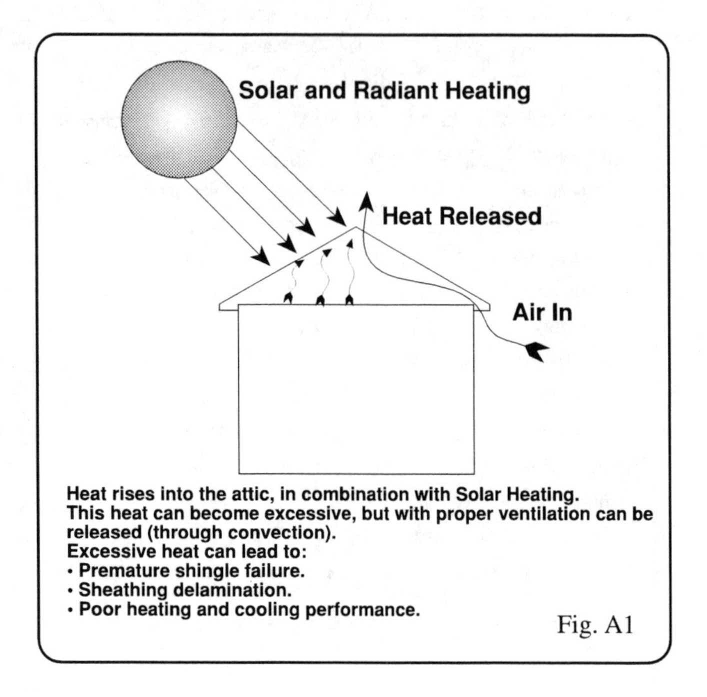

Fig. A1

Effects of poor ventilation may be apparent by mildew or moisture.

Heat may also lead to the plywood sheathing failing and even delaminating. Extreme heat can also become a fire hazard with certain stored combustibles. With darker colored shingles the situation is compounded, because they absorb and transfer more heat into the attic which can shorten the shingles' life span. When the attic is subjected to these conditions, shingles can actually bake and begin to curl; a sign of premature shingle failure.

With proper ventilation air is allowed to circulate, drawing in cooler outside air, and allowing the hotter air in the attic to

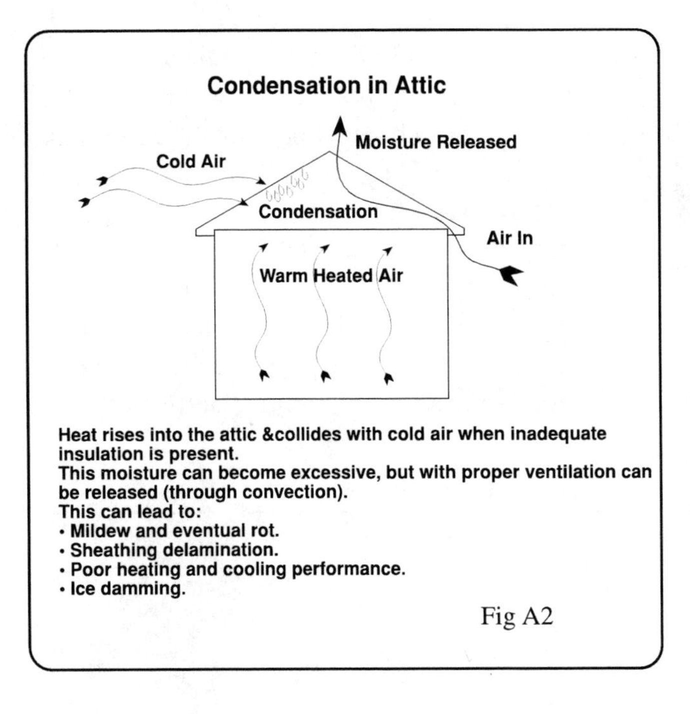

Fig A2

rise and escape. This can be accelerated with a powered ventilator.

In colder temperatures a dew point is created in the attic. (See Fig. 2A) Hot air rising through the ceiling (from the heat in the house) escapes, and collides with the cold air outside, forming condensation on the sheathing. The attic can become very damp, and with these conditions mildew and fungus can form, which can then lead to rot. In some cases the entire attic, including the rafters and sheathing, may be covered with a black mildew or fungus. Mildew can be killed with a mild bleach and water solution, and many times removing the moisture will cause the mildew to die. However, the problem will come back if proper steps are not taken to correct the situation and permanently remove the moisture.

INSULATION & VENTILATION

There are several solutions to chose from when revising poor attic conditions. First check for proper attic insulation. This is the first line of a thermal break for your house. Proper insulation can reduce the chance of a dew point occurring in the attic in colder temperatures. In hot weather, this insulation barrier can keep the hot attic from causing a dew point in the house due to the air conditioner. It helps to keep the heat from building up in the living space as well. You will want to make sure you have the proper amount of insulation in the attic for your region. The amount of insulation and how well it buffers determines the "R-value" and this will help your heating and cooling system's efficiency. If there is sheathing down for storage, be sure it isn't compressing the insulation and lowering the R-value.

Improper insulation can also help promote ice damming. (See Fig. 3A) Heat rising into the attic can cause the ice to perpetually thaw and refreeze, allowing the ice to creep under the shingles. Once this occurs, a leak may appear at the soffit or ceiling. Ice damming can also occur above skylights, chimneys, etc.

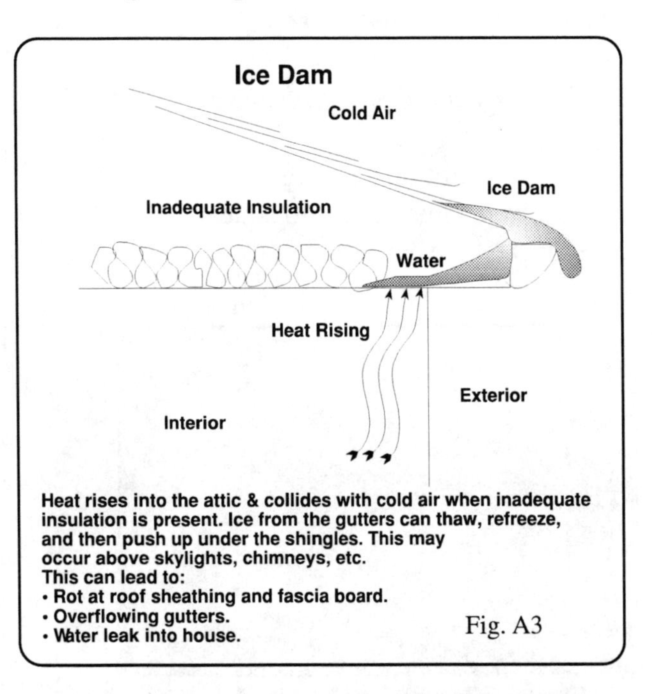

Fig. A3

Verify insulation is not obstructing soffit ventilation.

Fans

One ventilation solution is to install a power fan; the manufacturer will specify how many cubic feet per unit. The fan will be activated by the thermostat in hot conditions, but unless you install a humidistat, the fan will be ineffective in colder weather. This is a good system to add onto an existing roof; installation is minimal, simply requiring cutting a hole and connecting the fan to electricity.

Continuous Ridge Vent

A continuous ridge vent is another effective method to vent the attic. This utilizes the highest part of the roof (ridge) so the heat will naturally rise and escape. The sheathing has a gap about 2" for the heat to escape. A ridge vent is more involved to add on an existing roof, and involves removing the caps and cutting a swath down the ridge. If the shingles are being removed completely, this is more viable solution. Ridge vents come in metal and plastics with a honeycomb to allow heat to escape. The plastic is then covered by shingle caps.

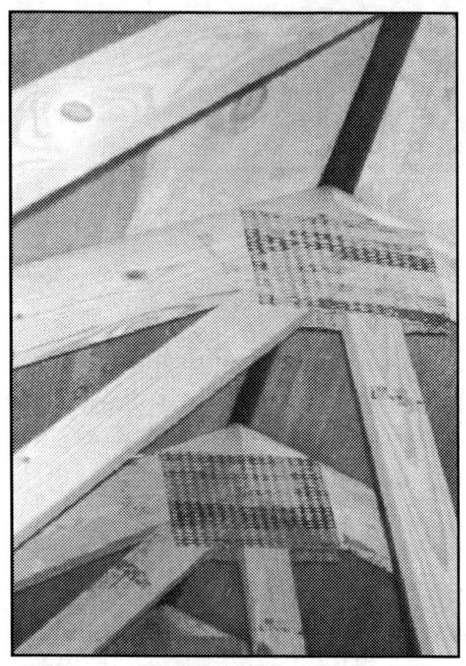

Ridge vents allow air convection. The aid of soffit vents allow heat and moisture to escape.

There are an array of other vents available:

- Low mount rectangular
- Turbine (susceptible to squeaking)
- Drip edge vent
- Apron vent
- Hip vent

Soffit vents allow cooler air to be drawn into attic space.

With any of these methods, the air needs a source to draw from, known as a soffit vent, which allows the air to circulate. Think of a straw with your finger over the end and water in it --- the water can't flow. The same thing applies to your attic. If you open the attic access, you will feel a rush of air as the attic draws the air up. If you do have soffit vents, make sure the vents are not blocked with

Look for water stains at all roof penetrations, including chimneys.

insulation. You can use a baffle (styrofoam or cardboard insert used to create air chase) designed for that purpose.

PLUMBING & ELECTRIC

While in the attic, look for any exposed plumbing supply lines that could be susceptible to freezing. Verify that all of the plumbing vents terminate outside. Inspect the electrical wiring for any hazardous conditions or open junction boxes. If there are any recessed lights in contact with insulation be sure that they are rated for contact. The wrong type of fixture in contact with insulation could be a fire hazard. Check any bath vent for proper venting to outside air. Often the vent will terminate into the attic or be blocked by insulation.

In general, bath vents should not terminate in the attic, as this adds high humidity to the attic space. A clothes dryer should not terminate into an attic either, for it poses a fire hazard as well as causing high humidity.

WILDLIFE

Look for signs of nesting. Inspect the gable vents for tight proper screens to prevent birds and rodents access. Check the gable wall for solid connection to the top plate and at the vent; sometimes the gable rafter is cut to install the vent and is not reinforced. Any other large gaps should be filled to prevent water or animal intrusion.

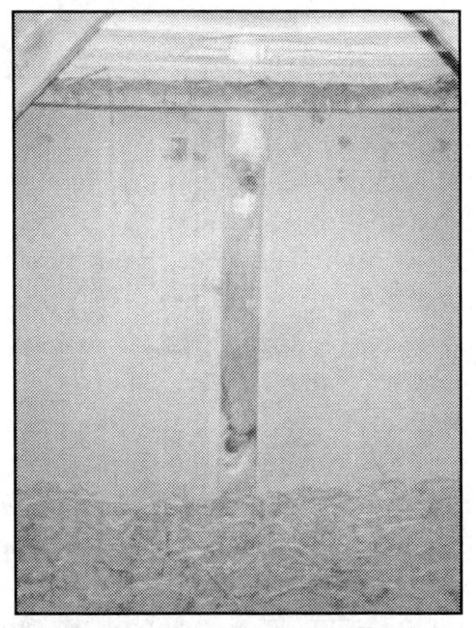

Torn gable vent screens allow animals to enter.

Modular roofs may be identified at double center connection.

STRUCTURAL PROBLEMS

Plywood By-Pass

- Excessive span or missing ply clips can allow the plywood to by-pass.
- May lead to premature failure in shingles.
- Sometimes mistaken for FRT failure.
- Recommend running 2x4's on edge, between rafters, and fastening sheathing to blocks at seams with construction adhesive and screws.

Sight down the trusses or rafters for any sagging or cracking. Look at the plywood or oriented strand board sheathing for by-pass. (See Fig. 4A) This occurs when the sheathing is missing ply-clips, which are used to keep the sheathing's edges aligned in between the trusses. If the ply clips are missing the sheathing can by-pass the piece next to it and transmit the imperfection through to the shingles. This can be repaired by nailing blocks in between the rafters and fastening the plywood to the blocks.

Probe the sheathing for any rot or signs of delamination. Sight the ridge pole and hip rafters if applicable. If there is severe sagging or crowning, further investigation may be required. Be sure collar ties or ceiling rafters are secure to prevent the walls from spreading.

In high wind areas you may want to look for rafter or truss tie-down straps connected to the walls.

Check that the attic stairs are properly connected and cut to the proper height. If there is a ridge vent, be sure a gap has been cut to the manufacturer's specifications. Some roofers will install the ridge vent when no gap was ever cut and forget to cut out the shingles. This will not allow the ridge vent to do a proper job. If you could not get on the roof this is an opportunity to see if the vent is properly fastened by pushing on it or looking for excessive daylight gaps. Look at the ends to see if the plugs are in.

Inspect any valleys or exposed flashings for possible leaks. Step flashing should have a backing; if not, the flashing can bend over into the attic and fail, letting water and pests in.

Rafters should seat tightly against ridge pole.

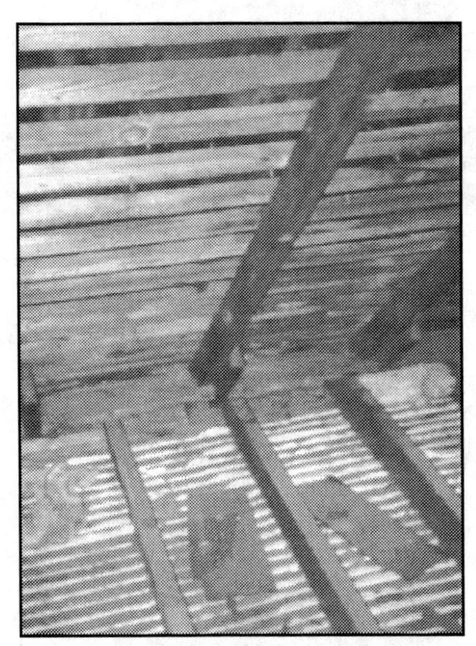

Skip sheathing was commonly used with cedar shingles as well as metal roofing. Note lack of insulation.

If you follow these suggestions, you will increase the life of your entire roof structure, and ensure your attic doesn't become a fire hazard. In addition, you will be helping the efficiency of your heating and cooling systems, and saving money.

Refer to Roof and Structure sections for other information regarding the attic.

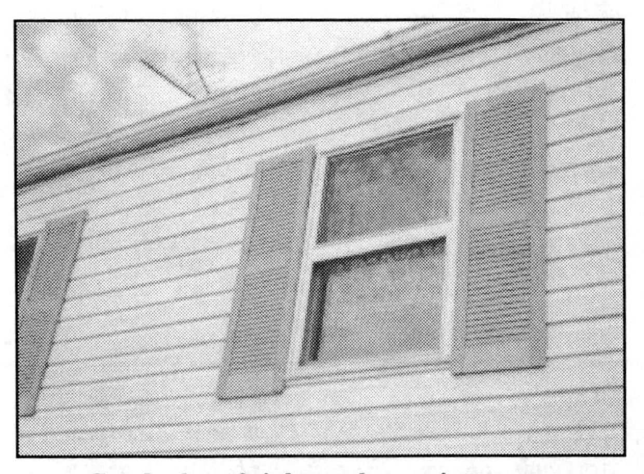

Gambrel roofs inherently restrict proper air circulation and may lead to premature shingle failure.

ATTIC & GARAGE CHECKLIST

1] **ACCESS** ☐Yes ☐Partial ☐***Access Restricted Due To;*** Blocked / Inadequate / Cathedral / Unsafe / None Found / High Insulation
☐**Pull Down Stairs** ☐**Scuttle** ☐**Walk Up** ☐**Eave** ☐**Problem** Pull Down Stairs Inadequately Installed, Damaged / Hazardous

2] **FRAME** ☐**Trusses** ☐16"oc ☐24"oc ☐Other ☐Wood ☐Beam(s) ☐Metal
☐**Rafters** ☐16"oc ☐24"oc ☐Other ☐Wood ☐Beam(s) ☐Metal
☐**Problem;** Sagging / Cracked / Excessive Spacing / Rot / Mildew / Cut Sections / Warped Sections / *Previous Signs Of Fire Damage and/or Repairs*
Ridge Pole ☐Yes ☐No ☐**Problem** Sagging / Settlement Gaps **Collar Ties** ☐Yes ☐**Problem** Recommended / Inadequate Nailing

3] **ROOF SHEATHING** ☐Plywood ☐Oriented Strand Board (OSB) ☐FRT ☐Solid Planking ☐Spaced Slats (Skip Sheathing)
☐**Problem;** Delaminating / Cracked / Mildew / Rot / Water Stains / Missing Plyclips / Plywood, Sheathing By-Passing Adjacent Sheet / Sagging / Ice

4] **MOISTURE, STAINS %**______ *Observed* ☐**Yes;** Plumbing Vent(s) / Ventilation / Skylight(s) / Gable / Eaves / Flashing / Chimney(s) / Valley(s)

5] **INSULATION** ☐Yes ☐N/V ☐**Problem;** None / Inadequate / Soffits Blocked / Missing In Areas / Animal Nesting / Wet, Damp

6] **ELECTRICAL** ☐Ok ☐**Problem;** Open Junction Boxes / Missing Junction Boxes / Sloppy, Abandoned Wires / Non-Rated Recessed Lights

7] **LIGHT(S)** ☐Yes ☐No ☐**Problem;** Non-Operational / Damaged Fixture / Improper Installation / Bulb Broken Off / Hazardous

8] **PLUMBING** ☐N/V ☐**Problem;** Visible Leaks / Susceptible To Freezing / Vent Terminates In Attic / Condensation / Visible Leaks

9] **VENTILATION** ☐Yes ☐Ridge ☐Soffit ☐Turbine ☐Gable Vents [☐**Power Vent** ☐*Not Tested*] ☐Other
☐**Problem;** None / Inadequate / No Screens / Screens Damaged / Blocked Soffit Vents / Ridge Inadequately Cut Open / Signs of Excessive Heat

•ADDITIONAL COMMENTS *•Attics with boxes, etc. could impede inspection and be a possible fire hazard. Attics that are not accessible to the Inspector should be reinspected.*

____

____

____

☐**Appears Adequate** ☐**Repairs Required** ☐**Further Evaluation Required** ☐**Blocked, Limited, No Access**

Garage

☐Yes ☐No ☐**Repairs Required** ☐**Further Evaluation Required** ☐**Amateur, Sloppy Workmanship**

1] **ATTACHED**[1] ☐Yes ☐No ***If Yes;* Fire Rated Door To House** ☐Yes ☐No ☐Unknown ☐*Recommended*

2] **OVERHEAD DOOR(S)** ☐Ok **# Of** ☐1 ☐2 ☐3 ☐4 ☐**Wood** ☐**Metal** ☐**Fiberglass** ☐**Composite**
☐**Problem(s);** Water Damage / Rot / Delaminating / Damaged Panel / Won't Stay Open, Closed / Dented / Sagging / Frame Failing, Cracked
Auto Opener[3] ☐Yes ☐No ☐Non-Operational ☐Older Unit **Lighted** ☐Yes ☐No ☐**Problem** Missing, Damaged Lens / Out

3] **SLAB / APRON** ☐**Ok** ☐**Problem; [Cracked** ☐Minor ☐Severe] / Covered[2] / Walls Blocked, Limited Access / Standing Water / Settling

•ADDITIONAL COMMENTS *[1]Unattached garage is not part of the standard inspection and is considered an additional building. [2] If slab is covered with personal items, the inspection may be blocked or impeded. [3]Auto safety reverse devices are recommended for an overhead garage door however they are not tested, since damage could result to the door.*

____

____

____

____

____

____

____

____

____

____

____

____

Section Four:

Plumbing

Supply & Waste

There are two basic sections of plumbing: supply and waste. There are different types of supply and waste lines and materials, but both can be on private source (well and septic) or public water and sewer. The supply and waste can be also a combination of one of each.

SUPPLY

The supply, or source, can arrive from a private well or public water. Public water usually creates little maintenance for the fixtures. Chlorine taste and odor are usually the only noticeable traits associated with public water. The pressure is regulated by the city.

Well water is supplied by a pump. There are two basic pumps: one pulls the water and the other pushes the water. An above ground pump is usually seen with shallow wells, whereas a submersible or foot pump is more efficient pushing up water from deeper wells. It is also a quiet pump since it is out of sight. The only drawback with a submersible pump is to repair or replace it, the pump has to be pulled from the well.

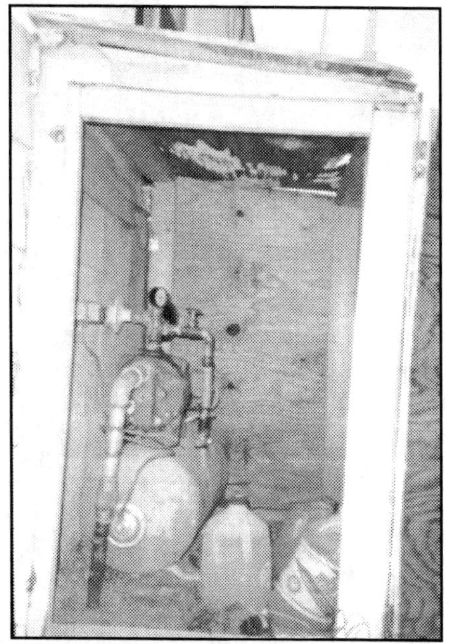

In colder climates well tanks and pumps exposed to the outside may be susceptible to freezing.

Water can be delivered in a plastic, galvanized steel, copper or other supply pipe where there should be a way to shut off the water supply, by means of a fairly easy access. (Polybutylene supplies were failing at the foundation due to break off, but now a sleeve has been added inside the pipe to help prevent this problem. See more about this under the "Polybutylene" section on the next page).

Once the water is drawn into the system it is routed into a storage tank. Older tanks are large galvanized reservoirs and are prone to rusting and collecting water contaminates, so routinely drain off these older tanks if you have one. Modern tanks use a bladder pressurized with air from the factory. As water enters the tank, the bladder compresses the air on the opposite side, and when the water is demanded, the water is pushed out under pressure. This cycle is regulated by a switch that is usually set to turn on and off at 20lbs and 40lbs or 30lbs and 50lbs. a cycle which varies depending on the size of the holding tank. When a tank short cycles, it may be over pressurized, not allowing the water to enter the tank. This is common even on new tanks if the installer never adjusted the proper air pressure in the tank. This can be verified by feeling if water is entering the tank, which can be done by lifting the tank if possible. Next, try tapping the sides of the tank, and listening for water to deaden the sound. Also, feel for cold on the sides of the tank, which could indicate the presence of cold water. Watch the switch for switching on and off --- with a submersible it is hard to hear the water pumping.

Another problem associated with lower pressure and short cycling is a water logged tank. This occurs when a bladder fails and water gets behind the bladder. In this case, the tank needs to be replaced. If a pump is allowed to continually run in this condition, its life will be greatly shortened.

Materials

Once the water has reached the tank it is ready to be supplied to the house. There are several materials used for supplies lines, and the following are some of the more common materials used. Generally, the main supply will be 3/4" and the supply branches will be 1/2". This may vary depending on codes, supply demand, and other factors.

• *Copper* is widely used around the country. There are two common thicknesses. The thinner is type M, which is easy to bend and which can be carried out with fewer fittings. It is also more susceptible to pitting and eventual failure in acidic water. The thicker gauge copper type L is recommended and less prone to failure.

Note that older solder used to connect the fittings may have contained lead. This can contaminate drinking water supplies, but replacing the fittings and solder can be very costly and impractical. If lead is found or suspected it is recommended, at a minimum, to let the water run a few minutes, before using. As the water sets in the lines all night, it builds with lead. Water running freely has less chance to collect lead.

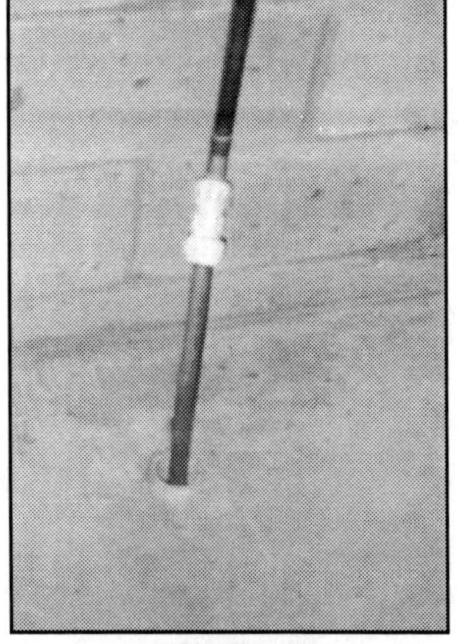

"Big Blue"
Polybutelyne main water supply has a history of failure.

• *PVC is* that familiar bright white plastic colored pipe. It is not generally recommended for use with drinking water and is not rated for use with hot water supplies.

• *CPVC* is an off white cream colored plastic pipe. It is accepted as a supply for potable water supplies and for use with hot water supplies.

• *Polybutylene Pipe* is a plastic developed for use as a plumbing product by Shell Oil Company. There are a few problems associated with polybutylene, associated with homes built from 1/1/78 to 7/31/95. In fact, most areas have banned the use of polybutylene. There are some class action law suits against the manufacturers as well as the installers and are trying to settle the damages. One has been settled for 950 million dollars and the fund will remain open until the year 2009 or 16 years after date of installation, whichever comes first.

WARNING!

Hot water supply should be delivered at 120˚F to save on energy and prevent scalding. According to the Consumer Product Safety Commission, it takes 5 minutes of continual exposure of 120˚F water to deliver a 3rd degree burn. However, at 130˚F the exposure time is reduced to 30 seconds, at 140˚F the time is 5 seconds, and at 150˚F the time is only 1.5 seconds.

Big Blue, Vanguard: (Claim must be within 11 years of installation)

The supply line from the outside is sometimes referred to as "Big Blue", and may be identified by its light blue color. This type of polybutylene is susceptible to shearing at the foundation wall due to ground settlement outside. One recommendation is to sleeve the the pipe through the wall for additional reinforcement. Another recommendation is to simply replace it. The plastic is also reported to break down due to elevated levels of chlorine;there has not been any established time frame in which this usually occurs.

Interior polybutylene with plastic connections may crack and allow for free flowing water.

Plumbing vent should be elevated ten inches above roof line. Inspect rubber boot for deterioration.

Qest, Vanguard: (Claim must be within 13 years (acetal fittings), 16 years (metal fittings) of installation)

Interior polybutylene is identified by its gray color. When this product first came out the ends were connected by metal barbed connectors, which were held in place with a copper crimped band. The metal fittings were overall adequate. Later, plastic barbed fittings were introduced. These fittings were prone to cracking off and failing, due to crimping of the copper or aluminum bands, and could allow water to free flow. The plastic fittings may develop hairline cracks from the pressure of the crimping and should be replaced with approved metal fittings.

In some areas highly acid water can deteriorate copper pipe in a relatively short time, or cause them to burst due to freezing. Polybutylene, on the other hand, is virtually unaffected by these conditions. Polybutylene has a good flow rate, is more impervious to freeze, and has fewer connections. We should be aware of its problems, but know its strengths as well.

Plumbing vent should extend above roof line and not terminate below eve or near window.

• _Galvanized Steel_ is found in older homes, often part of a system which has been otherwise updated. When the galvanized supply or branch piping is before the newer copper branch supplies, the system supply and pressure will be dictated by that section. As corrosion builds up inside the walls of the pipe, the supply is choked down, affecting the pressure. Typically sections of the pipe may need to be replaced. A 3/4" pipe's walls will commonly be choked down to the opening width of a pencil. Well water which has a mineral and iron content may accelerate this process.

If suffering from low pressure problems, the pipe can actually be cleaned out, either by high pressure or an auger. Either may cause damage to the supply pipes, and are not guaranteed to restore the water pressure. The best method is to replace the old pipe with new piping.

"S" trap - older style for drain prone to gurgling due to inadequate venting.

All supply lines and risers should be properly supported and strapped, and any sagging or vibrating supplies should be secured properly. Mixed metal straps may corrode and fail. To prevent this from occurring, the straps should be replaced with compatible straps.

The temperature can be turned down at the hot water tank, with relative ease.

WASTE

There are several materials used for waste plumbing:

- PVC, ABS (usually schedule 40)
- Copper
- Cast Iron
- Ceramic
- Galvanized

There are a few common components that make up a drainage system:

• *Vent & Stack*: allows pressure relief on the drainage system, and inhibits possible siphonage of the traps. The vent can be a dry or wet/dry stack (shared with a drain). A toilet should have a 3" or greater vent size. Any horizontal runs should angle slightly to allow any water to drain out. The stack should penetrate the roof and stick up approximately 10" above the roof. If the vent is too short, snow, leaves, and other objects may cover and impede proper drainage and ventilation.

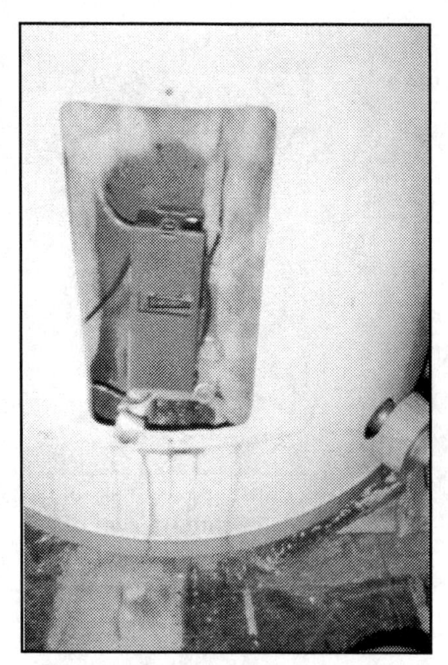

Electric water heaters may show signs of rust or leakage at elements drain valves or base of jacket.

The vent should not terminate under an overhang, in an attic, or within 6' of a window or door (and should extend at least 2' above either location). Some areas allow a jiffy vent (vent with a check valve to prohibit sewer gas from escaping) near the fixture, however, most areas do not allow this type of vent. This vent may fail, and then allow sewer gas to enter the house. Check with local building codes.

• *Trap(s)*: are used to create a water seal in a U shaped drain that prevents sewer gas from backing into the house. A "P" trap is the standard today, and is identified by the horizontal section draining back into the wall. This section should intersect at a vent.

"S" traps drain down into the floor and are generally inadequately vented. In some cases siphon action can pull the water from an "S" trap, which could allow sewer gas to enter the house. If a drain is slow or gurgles, poor venting may be the cause. If a sewer odor is present, poor venting or a dry trap is usually the cause. When a bathroom, floor drain, or other remote fixture drain is seldom used, the trap seal (water) will evaporate. Once this occurs, sewer gas can enter the house. These traps should be primed regularly. There are self priming traps available as well.

A fixture should never extend into a basin, This could allow waste water to contaminate a potable water supply. Areas such as a laundry sink, tub, sink, hose bibs, etc., are potential problem areas.

• *Clean-out:* a plug (generally screwed in) that allows access and cleaning of the interior of a pipe.

• *Drain*: takes all of the sewage and other water waste from the building to a legal point of disposal, which may be either public or private. The main drain should be at least a 4" diameter pipe. The fixtures will vary as to the size of the drain required. Generally, in a residential building, a water closet (toilet) will have a 3" soil stack and drain, and other fixtures will generally have a 1.5"-2" stack and drain.

The drains should be at a uniform sloping pitch to the point of disposal. The drain should be properly supported and not sagging at any point.

In areas which won't perk a raised mound system might be a solution for proper septic drainage.

Waste is passed out to a legal point of disposal. It is illegal for waste water to pass from the house (including laundry water and

Gas water heater flue should be secured to hood, no visible corrosion, and lead away with male to female connections.

utility sink) to drain onto the open ground, to a body of water, or to a non-contained area. The waste must drain into:

- Private holding tank
- Private septic system
- Private treatment system
- Public treatment (city sewer)

Most private septic systems are comprised of a septic tank in which the waste is collected. Most septic tanks are now concrete, however, older tanks may be metal and prone to rusting and leaking.

The solids collect in the bottom and the liquid rises to the top and flow to the drain-field. The drain-field is usually comprised of perforated drain tile buried in a gravel bed, covered with a filter cloth to prevent dirt from washing into the field and choking off the flow of water. The quantity and length of these fields are in direct proportion to the number of bathrooms and other plumbing fixtures, the number of residents in the house, and absorption rate of the soil. If more than one field is deemed necessary, a distribution box may need to be installed. This divides the flow between the different fields. A perk test is carried out before the system is installed to determine if the soil will handle liquid saturation. If it is determined the soil will not handle the load, fill dirt may have to be brought in, or it may warrant the installation of a mound system above grade. Sometimes a private system will not work.

In either case, the septic system should be periodically checked, though a proper system should need little maintenance. The bacteria should dissolve the solids and create a cycle. However, if the cycle is disrupted by chemicals introduced into the system, the sludge at the bottom of the tank may solidify and buildup, eventually choking off the tank. Routinely pumping the tank could prevent any problems.

FREEZING PROBLEMS

Any supply or drain pipes in colder exterior walls, crawl spaces, attics, garage, or drafty areas may be susceptible to freezing. In these cases foam insulation should be installed to help prevent freezing. Electrical heating tape can become a fire hazard and is not recommended. A separate heat source may be necessary in some extreme cases.

If you live in the colder regions and you leave the house for a few days or more, you may want to consider draining the system and winterizing your plumbing. This is accomplished by turning off the main water supply, and then opening all of the water fixtures to allow the water to bleed from the system. If you are on well water you should consider draining the water tank as well; there is a hose bib connection at the bottom of the tank. Next you should drain the hot water heater from the valve at the bottom of the tank. Be sure to turn off the hot water heater before draining the tank.

Exterior hose bibs should be winterized to avoid freezing.

If you know you will be away from your home for some time, you may also want to protect the drain system. The toilets should be filled and cycled with a type of antifreeze that is non-toxic. All of the traps in the house should also be filled with antifreeze. RV and boat antifreeze work well for these applications.

If you have a boiler type furnace and the house is going to be vacant for an extended time in the winter, you may want to consider using a special antifreeze in the system. This way, if the heat ever fails, the boiler will not freeze.

It is also recommended that an exterior spa or hot tub be drained when the house is not occupied for any length of time in the winter. When the power is interrupted and the water in the tub does not freeze, the water in the pump may. Once the power is restored the pump can burn up if still frozen. Often exterior units are not covered under a home owner's policy.

Exterior Hose Bibs & Freezing Temperatures

The hose connections are referred to as the hose bibs. These standard bibs are commonly not freeze-proof and need to be bled in the winter. To properly winterize your exterior hose bib, a shutoff valve must be located inside the house and turned off. Next, the outside hose bib must be opened to drain the remaining water in the pipe.

"Freeze-proof" hosebibs are not always 100% freeze proof; they have a shutoff plunger which shuts off the water several inches inside the wall within the heated home space.

Some hose bibs have an anti-siphon built in to prevent outside water from being drawn into the system. This keeps outside contaminates from being drawn into the water supply.

EIGHT EASY STEPS TOWARDS WINTERIZATION

When winter comes, colder weather and the effects of ice, snow and other elements on a structure need to be addressed.

1. Caulking around windows, doors, trim, and gaps in siding can cut down on drafts and water infiltration. Putties are available for larger holes and exterior wall penetrations such as plumbing and electrical lines.

2. Seal asphalt driveways before the ground freezes. Water may infiltrate small cracks, freeze, expand, and cause great damage. For larger cracks, a thick sealant in a caulking tube should be used. Cracks in concrete, brick or block should also be filled or sealed to prevent further damage. This includes sidewalks, foundations, driveways, veneers, chimneys and chimney caps. Small cracks can be filled with a polyurethane caulk or a masonry sealer. Larger cracks may be filled with hydraulic cement. This is a waterproof cement which will create a permanent bond. There are hydraulic cements available for coating concrete such as a concrete cap on a chimney. This prevents water from being absorbed into the concrete which will reduce the chance of freezing and cracking.

3. Prepare for ice damming. Ice damming occurs where ice is allowed to build up, and under force pushes against or under another area. Gutters are probably the most common place where this occurs. Snow melts from the roof and freezes in the gutter, and as the ice mounds in the gutter, the ice may push under the shingles and onto the roof sheathing (plywood, OSB, etc.) and remain there until it melts. This can cause a leak in your ceiling or soffit overhang. If this condition is allowed to continue, rot can occur to the sheathing, frame, fascia and wooden soffit.

There are preventative steps that can be taken, and a drip edge is probably the easiest and least expensive. F-Style drip edge is a metal strip that comes pre-bent in 10-12' sections. It goes under the shingles about 2", overhangs the edge of the roof into the gutter about 3/4", comes down over the fascia about 1" and has a lip for a positive run off. This creates a seal to the edge of the roof and greatly reduces the chance of ice damming or water damage. Other methods of prevention include custom bending larger widths of flashing for greater coverage protection. The state of the art is a rolled rubber type membrane that is extremely sticky on both sides and bonds to the sheathing as well as to the shingles. This has to be applied when the shingles are removed and the sheathing is exposed.

Other areas to check for ice damming are above skylights, chimneys, valleys and adjacent walls. Preventing or repairing these areas may be as simple as sealing the vulnerable area with roofing cement. However, you may want to have a professional come out to repair the roof. Some things are better left to a qualified individual.

4. <u>Drain</u> plumbing and be sure to bleed the water out of the hose bibs, especially if they are not freeze-proof.

5. <u>Insulate</u> pipes exposed to cold drafts. There are wrapping fiberglass styles or soft foam cell sections with a slit which you slide over the pipe. These all are in short supply once the first big freeze hits, so do not put it off until then. Also wrap your water tank (if applicable) and hot water heater.

Proper insulation throughout the house is recommended. Check with your local building codes to verify proper R-Value for attic, walls and floor. The attic is usually the easiest to access and is good place for your heat to escape. Covered walls can have additional insulation blown in. Exposed floor joists in a crawlspace should be insulated, and basement walls above grade may need insulation as well. Remember, in some cases, once the floor is insulated the heat source that once kept your plumbing from freezing will no longer warm the pipes.

6. <u>Plug</u> exterior wall outlets which allow drafts to come through from outside. There are gaskets available to put behind the cover plates.

7. <u>Install</u> storm windows, replacement windows, or put up plastic to help prevent heat loss, condensation or to stop ice from accumulating on the interior of windows.

8. <u>Service</u> your fireplaces or wood stoves and have chimneys cleaned to ensure proper burning and safety. Maintenance will also help to get maximum life from the units. Clean and service your heating system to ensure maximum efficiency and safety. Change the filter and have the blower cleaned when applicable.

PLUMBING CHECKLIST

1] **SUPPLY** ☐**Public Water** ☐**Well Water** ☐**Private** ☐**Water Treatment** *(Only Inspected For Visible Leaks)*

a] **Pressure Tank** ____ / ____ **lbs** ☐**Approximate GPM Flow**_____ *(Optional Test)* ☐***Low Pressure***[1] *(Have Checked by Plumber or Well Company)*

☐**Problem**[1]; Water Logged / Not Cycling Properly / Visible Leaks / Rusted / Gauge Stuck, Not Operating / Condensation on Tank / **Older Unit**

b] **Supply Line** ☐Copper ☐Plastic ☐Galvanized ☐*Polybutylene*[2] ***(Is susceptible to failure from settlement)*** ☐Other________

☐**Problem**[1]; **[Shut Off Valve**; None / Not Located / Valve Leaks] / Susceptible to Freeze / Sweating / *Polybutylene*[2] / Inadequate Size

c] **Interior** ☐Copper ☐PVC *(Not rated for hot water)* ☐CPVC ☐*Polybutylene*[2] *(Is susceptible to failure)* ☐Galvanized

☐**Problem**[1]; Visible Leaks / Mixed Metals / Susceptible To Freeze / Hanging Loose / Corrosion at Connections /*Amateur, Sloppy Workmanship*[1]

[***Corrosion;*** Moderate, Extensive] / Pin Hole Leak(s) / Signs of Previous Leaks / Green Corrosion *(May be from acidic water and may lead to plumbing failure.)*

d] **Hose Bibs** ☐Yes ☐Not Located **Freeze Proof;** ☐Yes ☐No ☐*Recommend Draining Hose Bib of Water In Winter*

☐**Problem**[1]; Drips / Shut Off or Not Functioning / Handle Missing / Handle Damaged / Not Winterized / At Grade*(Susceptible to backventing supply)*

2] **WASTE** ☐Public Sewer ☐Septic **Material;** ☐Plastic ☐Cast Iron ☐Galvanized ☐Copper ☐Ceramic ☐Lead

☐**Problem**[1]; Visible Leaks / Improper Venting / Negative Slope / Sewage Odor / Corroded / Cracked / Rusted / *Amateur, Sloppy Workmanship*[1]

a] **Traps;** ☐Metal ☐Plastic ☐Other ☐**P-Traps** *(Modern)* ☐**S-Traps** *(Older style, susceptible to gurgling, siphonage and slow drainage)*

☐**Problem**[1]; Visible Leaks / Improper Venting / Failing / Loose Connection / Corroded *(Replacement Recommended)* / *Amateur, Sloppy Workmanship*[1]

3] **UTILITY SINK** ☐Yes ☐**Plastic** ☐**Concrete** ☐**Stainless** ☐**Other**________________

☐**Problem**[1]; ***Visible Leaks At;*** Valve Stems, Spigot, Supply Line, Drain, Basin / Damaged / Spigot Below Water Level /*Sloppy Installation*[1]

4] **WATER HEATER** ☐Yes *(Conventional)* ☐Supply Demand **MFD**______ **Fuel Type;** ☐**Electric** ☐**Gas, Propane** ☐**Oil** ☐**Solar**

☐**Problem;** Not Heating *(Turned Off)* / Too Hot *(Set to 125°F)* / [***Leak, Rust At;*** Tank, Element, Valve] / No Temperature Pressure Relief Valve *(TPR)*

Not Accessible / TPR Valve Not Directed Down / No Overflow Pan / Gas Odor / Burner Dirty / Flue Missing, Loose, Corroded, Failing /***Older Unit***

•**ADDITIONAL COMMENTS** *Inspection is limited to visible plumbing and cannot be accountable for concealed conditions or susceptibility to freezing. [1]Further evaluation and/or repairs recommended by a licensed plumber. [2]Polybutylene plumbing is susceptible to the plastic connections failing and/or shearing at the foundation wall which could result in water free-flowing. Older plumbing may allow lead to leach Into the water. Older galvanized plumbing may have restricted water flow. Electric water heaters have an 8-12 year life expectancy.*

WATER ☐**On** ☐**Off** *(Could Not Evaluate)*[1] ☐***Further Evaluation and Repairs Required by Licensed Plumber*** [1]

Bathrooms

Bathrooms are built and designed with many different styles and components. Some of these components are prone to water related problems, which may result in damage. With proper installation methods and routine maintenance, the likelihood of serious problems, specifically water damage, may be averted.

TOILETS

Toilets come in a few different styles, including standard two piece, one piece, low volume pressurized, and commercial (high pressure) types.

A standard toilet uses a higher volume reservoir which is mounted above the bowl and forces the waste out of the bowl, through a siphon action. The water enters the bowl from the tank, through a series of holes under the rim of the toilet, and through the jet. This incoming water creates a siphon, that discharges the bowl water and contents. The tank and bowl are supplied with fresh water, and the internal trap is filled to create a water seal. (See Fig. 1B)

Fig. 1B

How a Toilet Works

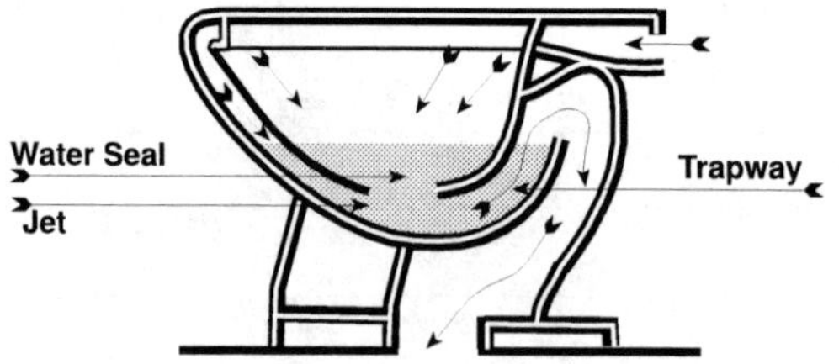

Water Seal

Jet

Trapway

A siphon is created from water rushing into the tank from the bowl.

Water enters the bowl through a series of holes around and under the rim and from the jet.

Inside the tank, inspect the condition of the flapper and float for proper operating condition and free movement. Well water with a high mineral content may cause the fill to become clogged occasionally. The flapper should seat over the tank opening, so water does not run out. These components are inexpensive and fairly easy to install.

Look at the tank/bowl connection for an even secure connection, and for no signs of leakage. A lower profile, one piece toilet will sometimes discharge at a slower rate. Modern toilets now use lower volumes of water.

The low volume residential toilet has a pressurized bladder inside. As water is supplied to the tank, it pushes against a diaphragm that is backed with a cushion of air. When the toilet is flushed,

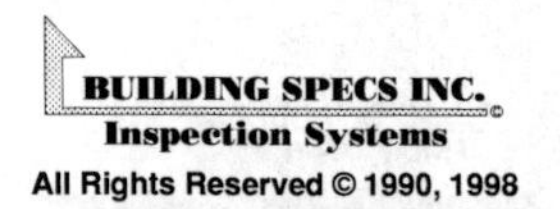

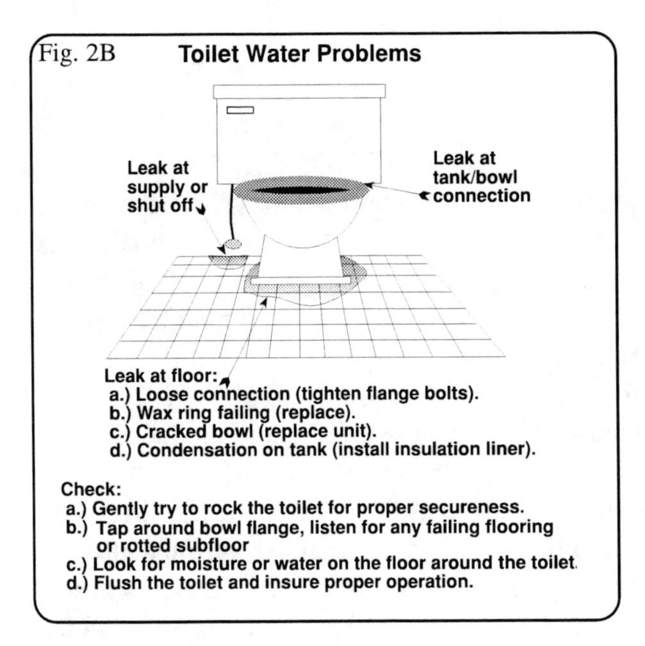

the smaller volume of water is released under pressure. If the bladder becomes damaged, it may become water logged and lose pressure.

When inspecting toilets, check for proper water levels in the bowl and tank as well as proper discharge (See Fig. 2B). If the toilet flushes too slowly and does not drain properly, it could mean a blockage in the drain, inadequate venting, clogged rim holes, or an inadequate water supply to the bowl. (The rim holes can be cleaned out with a coat hanger.) If the water continually runs, the float or flapper may be stuck.

Next, grab the toilet bowl, and firmly but gently, try rocking the toilet. You must do this gently, because if the toilet is too loose you could end up pretty wet. If the toilet is loose, the flange may have to be tightened against the wax ring. If the bowl is extremely loose, a new wax seal should be installed. The toilet should not be caulked into place, for this can allow a leak to go unnoticed. In this event, damage could occur to the floor structure.

When the toilet is removed, a closer examination of the floor is in order. Kneel down on the floor, and feel for moisture on the floor around the toilet. Then tap the flooring around the toilet, and look for loose tiles, damaged flooring or weak areas that may indicate rot. If you suspect rot below vinyl flooring, try inserting an awl behind the toilet (out of sight), into the subfloor. Inspect the toilet for cracks at the flange or in the tank or bowl. Evaluate the overall condition of the toilet and seat. The date in which the toilet was built is usually on the underside of the tank lid or inside on the tank. This may also give a clue to the house's age, or when the bathroom was remodeled.

Condensation may form on the tank due to cold water. There are insulation liners available to prevent this from occurring.

If there is a bidet present, check it for proper function, and do a similar evaluation on the unit as mentioned for toilets.

Check toilet for proper flushing operation and water level, or leakage at fill tube or flapper.

SINKS

Sinks come in a range of styles:

- Built-in
- Pedestal
- Wall mounted

Sinks also vary in materials used:

- Cultured marble
- Corian
- Porcelain
- Tile
- Enamel
- Stainless Steel

Turn the sink faucet on, and look for any leaks around the faucet or knobs. Make sure the aerator is in place. Check the stopper for proper function, or note if it is missing. Look at the condition of the sink itself for any cracks or defects that could cause a leak. Cosmetic defects may be noted as well. Gently try to lift on the counter --- it should be secure. A loose counter top can be secured with a silicone bead between the cabinet and counter. If the sink is wall mounted, check to see if the sink is securely fastened to the wall. If there are shut-off valves, check for proper operation. Make sure the sink drains adequately. Note: a slow drain could be the result of a clog or inadequate venting.

Look under the sink while the water is still running, and check for leaks at the drain and supplies. Inspect the trap to see if it is in good condition. If the trap is metal, tap it with your fingers and listen for any "thinning" sounds. *Gently* squeezing these areas can indicate signs of metal fatigue. Note the presence of water stains in the base of the cabinet (if applicable) which could reveal a past or persistent leak. In some cases, the cabinet base may have been completely water damaged.

TUBS

Tubs come in a range of styles:

- Built-in (standard)
- Whirlpool (recessed)
- Freestanding (claw foot)
- Custom (tile)

The materials can vary as well:

- Fiberglass
- Enamalized
- Plastic
- Tile

Tub and Shower Potential Water Damage

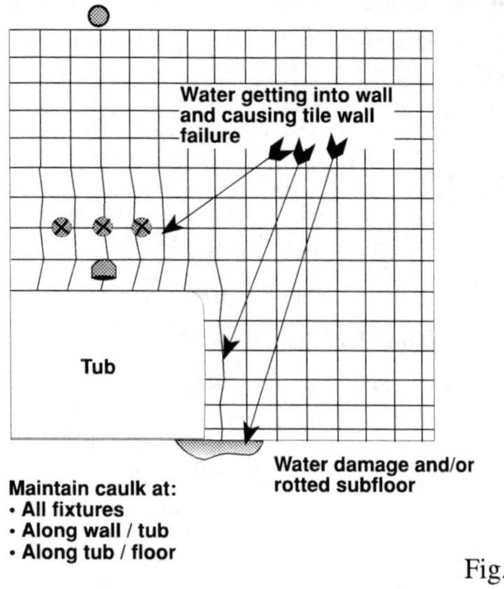

Water getting into wall and causing tile wall failure

Tub

Water damage and/or rotted subfloor

Maintain caulk at:
- All fixtures
- Along wall / tub
- Along tub / floor

Fig. 3B

When inspecting the tub, be sure it is on a structurally adequate floor system which will support the tub's weight when full. Be sure the tub is in good condition, with no holes or imperfections that could cause a leak. Check the condition of the plumbing fixtures for proper operation, including faucets, mixer, drain, and shower head. Be sure the fixtures are secure and not loose in the wall. Examine the tub's spigot which is prone to corrosion on the bottom. Turn on the shower diverter to see if the water is fully diverted from the tub's spigot. Inspect for proper drainage and function of the stopper.

Note if there is a plumbing access panel. If there is, remove the access panel to see if there is any evidence of water damage past or present, and check that the overflow is properly connected to the drain. If there is no access panel, servicing the tub in the future may include unexpected demolition of a finished area. If the panel is painted in place you may not want to remove it so as not to damage any surfaces. Note that if the supply lines are on an exterior wall you could have no access and could have "cold wall installation" (susceptible to freeze).

To check for proper water pressure, turn on all of the water in the bathroom, and flush the toilet, noting any substantial drop in water pressure. If the bathroom is on a second or third floor, compare it with the water pressure on the lower levels. Also, if the bathroom is far from the hot water heater, note how long the water takes to get hot. In extreme cases, booster pumps may be installed.

Tubs can have a combined shower stall built in as well. In this case, the tub needs walls which are protected from water.

There are several materials used:
- Fiberglass (one piece).
- Fiberglass (sectional walls).
- Ceramic tile.
- Marlite (masonite backed water resistant).
- High impact resistant plastics.

It is imperative that the seals at the tub and walls and around the faucet and handles are maintained. This will help to prevent water intrusion and damage to the wallboard, framing, subfloor, and ceiling below (if applicable) from occurring. Note any water running around the top of the tub which could run out onto the adjacent wall and cause damage to the wall and floor.

Tap the wall and push firmly but gently on the walls, paying close attention to the spigot areas (See Fig. 3B). Weak or spongy areas could mean water damage. If there are tile walls, tap them with your fingers, and listen for "clicking" sounds, which could indicate loose tiles. Inspect the condition of the grout closely for missing or failing grout. Run your finger nail through the joint to see if the grout is sandy or is failing. In some cases, a poor attempt at covering up previously failing grout may be evident. These repairs may include using grout that is wider than normal and which scrapes away easily from the lower layer.

Tap the floor along the tub and look for any failing tiles or signs of rot to the subfloor. Check adjacent walls outside the tub for water damage. Water can run across the top of the tub, or onto the connecting seam of some fiberglass surround walls. Also note any soap dishes or grab bars in the tub for secure installation.

If there is a whirlpool, inspect the shell and components as you would a standard tub. Then fill the whirlpool with water and test the cycles of the unit. Do not run the unit dry and be sure to aim the jets down, so water does not shoot out of the tub. The controls are operated by air or low voltage. Open the air venturies for full pressure. Locate access for pump and mortar. This is sometimes under

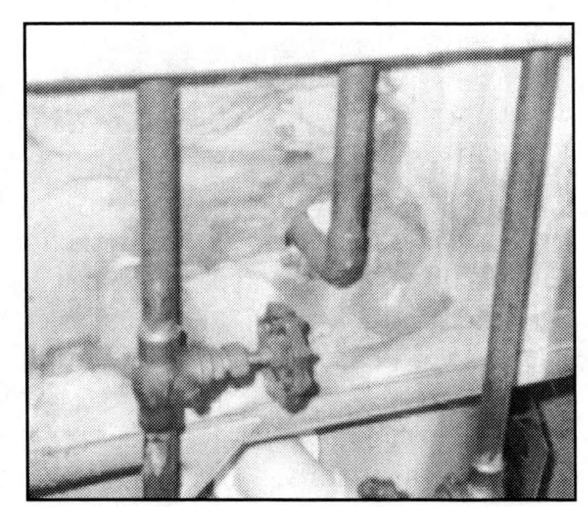

Evidence of water damage at shower/tub fixtures due to lack of caulking.

the unit behind a tile panel which is caulked in place.

SHOWER STALLS

Shower stalls may also be independent of the tub and can be:
- Prefab unit (floor and walls)
- Prefab floor (site built walls, tile, etc)
- Custom built unit

The shower may have glass doors, a curtain, or be large enough to be a walk-in. If there are glass doors, check their condition to insure they are sealed properly, and not susceptible to leaking. They should be made from tempered glass and operate and seal properly. Inspect the rest of the shower, as mentioned in the section on tubs.

FLOORS

Flooring in the bathroom may consist of:
- Ceramic tile
- Linoleum tile
- Carpet
- Sheet goods
- Hardwood
- Slate, Quarry tile, Brick
- Marble, Granite

Refer to the room survey on inspecting floors.

Problem areas which need close attention

- Toilets and Bidets, due to failing wax ring seal. Tap around the floor and look for spongy flooring. You may also test with a pinless moisture tester, looking for moisture below the flooring. The flooring should go under the unit.
- Along outside tub areas and shower floors.

Check for the presence and condition of the marble threshold. This type of sill is susceptible to cracking. The floor should be designed to hold any water that could seep through into the subfloor. Check all peripheral devices for proper attachment, such as towel bars, grab bars, toilet paper holders, soap dishes, toothbrush holders, etc. Note the condition of any existing mirror.

See that all lights function properly. Look for ground fault interrupter devices near all water sources, and check them for proper function. If a light is directly over a tub or shower, it should be of a moisture resistant type. These usually come with a gasket at the lens, to seal out steam and condensation. If there is a ceiling fan check for proper operation, which includes a quiet running motor and proper venting to outside air (not terminating in an attic). In bathrooms with a window, vents are commonly not installed.

Some bathrooms will have a separate heat lamp on a timer. Turn it on and check for proper operation. Identify the type of heat source. Note: If a register is in the floor, near the toilet, an overflow could run into the ductwork. The toilet's water could work its way into the heating system. If this occurs, the odor may be difficult to eliminate.

BATHROOM CHECKLIST

1] **FULL BATH(S)** # of______ **HALF BATH(S)** # of______ **UNFINISHED** # of______ **ROUGHED IN** # of______

2] **TOILET(S)** ☐Ok ☐**Problem;** Loose / ***Leak Observed At;*** Base, Water Supply, Tank Connection / Floor Rotted, Weak / Water Damage [Bowl, Tank; Cracked, Damage] / Low Pressure / Constantly Runs / Not Operating Properly / Seat Damaged, Loose / *Replace Wax Seal & Secure*

3] **BIDET** ☐Ok ☐**Problem;** Loose / ***Leak Observed At;*** Base, Water Supply, Tank Connection / Floor Rotted, Weak / Water Damage [Bowl, Tank; Cracked, Damage] / Low Pressure / Constantly Runs / Not Operating Properly / Seat Damaged, Loose / *Replace Wax Seal & Secure*

4] **SINK(S)** ☐Ok ☐**Built In** ☐**Wall Mount** ☐**Pedestal** ☐**Problem;** Damaged, Cracked / Heavy Wear / Loose / Leaks

a] **Faucet** ☐Ok ☐**Problem;** Drips / Leaks / Loose / Poor Condition / Low Pressure / Sloppy Flow / No Hot, Cold Water / No Water

b] **Drain(s)** ☐Ok ☐**Problem;** Clogged, Slow / Leaks / Loose / Amateur, Sloppy Installation / Failing / Corroded / Stopper Missing, Leaks

c] **Cabinet** ☐Ok ☐**Problem;** Water Stains, Damage / Heavy Wear / Damaged / Filled, Limited Access / Amateur, Sloppy Installation

5] **TUB** ☐Ok ☐**Cast Iron, Steel** ☐**Fiberglass** ☐**Whirlpool;** Not Operational / Not Tested / *No Visible Access To Pump*

☐**Problem;** Cracked / Chipped / Heavy Wear / Loose / Water Not Completely Draining / Amateur, Sloppy Installation / Rusted / Low Quality Unit

a] **Walls** ☐Ok ☐**Ceramic, Marble Tile** ☐**Fiberglass** ☐**Coated Masonite** ☐**Wood** ☐**Wallboard** ☐**Plastic**

☐**Problem;** Loose Tiles / Grout Failing /*Caulk Missing, Failing at Fixtures* / Water Damage / Walls Poor Condition / Amateur, Sloppy Installation

b] **Faucet** ☐Ok ☐**Problem;** Spigot Drips / Mixer Valve(s) Leaks / Loose Fixtures / Poor Condition / Low Pressure / Sloppy Flow No Hot, Cold Water / Not Diverting Water Completely / Spigot Corroded / No Access Panel / Spigot Below Water Line *(Possible Backventing)*

c] **Drain(s)** ☐Ok ☐**Problem;** Clogged, Slow / Leaks / Failing / Corroded / Stopper Missing, Leaks / Amateur, Sloppy Installation

6] **SHOWER** ☐Ok ☐**In Tub** ☐**Stall** ☐**Custom** **Shower Pan;** ☐**Fiberglass** ☐**Ceramic Built Up** ☐**Other**

a] **Walls** ☐Ok ☐**Ceramic, Marble Tile** ☐**Fiberglass** ☐**Coated Masonite** ☐**Wood** ☐**Wallboard** ☐**Plastic**

☐**Problem;** Loose Tiles / Grout Failing /*Caulk Missing, Failing at Fixtures* / Water Damage / Walls Poor Condition / Amateur, Sloppy Installation

b] **Water Damage At;** Wall Fixtures / Wall Penetrations / Widow Sill / Wall Bordering Tiles / Adjacent Floor / Top of Shower / Soap Holder

c] **Glass Doors** ☐Ok ☐**Problem;** Cracked / Damaged / Leaking / Poor Condition / Recommended / Seals Failing / Loose Frames

7] **VENTILATION** ☐Ok ☐**Problem;** Not Operational / Noisy / Inadequate / Mildew, Moisture Problem / Improper Discharge / Recommended

8] **HEAT** ☐Ok ☐**Separate Unit;** Wall Mount / Baseboard / Ceiling ☐**Problem;** Not Operational / Noisy / None / Recommended

9] **FLOOR** ☐Ok ☐**Sheet Goods** ☐**Linoleum Tile** ☐**Ceramic, Marble Tile** ☐**Hardwood** ☐**Carpet**

☐**Problem;** Water Damage / Grout Failing / Torn / Rot / Cracked Tiles / Heavy Wear / Loose Tiles / Caulk Failing / Sagging / Curling / Cupping

10] **LIGHTS** ☐Ok ☐**Problem;** Not Operational / Damaged / Cracked Lens / Melted Lens / None / Hazardous / Amateur, Sloppy Installation

11] **RECEPTACLE** ☐Ok ☐**Problem;** Not Operational / GFCI Not Functioning Properly, Not Tripping Properly / None Located / No, Open Ground False Ground / Reverse Polarity / No Ground Fault Interrupter (GFCI) *(Recommended)* / Hazardous Condition / Amateur, Sloppy Installation

•**ADDITIONAL COMMENTS** *'A ground fault interrupter is recommended within 6' of all water sources to prevent electrical shock. Caulking must be maintained at all shower and tub wall fixtures and penetrations to help prevent water damage to the wall and adjacent floor. Multiple Bathrooms are incorporated into this one section. Inspection does not include towel bars, etc.*

Powder Room: ______________________________

____

Master: ______________________________

____

First Floor Hall: ______________________________

____

Second Floor Hall: ______________________________

____

Basement: ______________________________

____

____

____

☐***Repairs Required*** ☐**Further Evaluation Required** ☐**Incomplete Installation of Components**

Section Five:

Heating & Cooling

Heading

There are many types of heating and cooling systems which may be utilized with each other. For example, a heat pump could have electric, oil, gas, or other type of back-ups. Oil and gas forced air can share ductwork with an air conditioner. Hydronic coils are used in the air handlers for heating along with a shared (split) air conditioning system. Any heating system may have an alternative solar system, either shared or separate. All of the systems require various degrees of maintenance.

Heating a building can use one or more principals:

- Direct (Woodstove)
- Exchange (Hot water baseboard)
- Passive (Window orientation to sun)
- Radiant (Hot water in floor)

HEATING SOURCE

There are several fuels, or sources, from which a heat gain is achieved:

- Oil (Hot water, Steam, Radiant, Forced Air)
- Gas - Either Natural or Propane (Hot Water, Steam, Radiant, Forced Air)
- Electrical (Elements (Resistance), Baseboard, Forced Air)
- Geothermal (Ground, Water Source)
- Solar - Passive or Active (Hot Water, Building Orientation, Forced Air)

Any system which involves a fuel that burns, such as oil, gas, propane, coal or wood, should be serviced annually. Any of these produce various levels of carbon monoxide. This is a colorless, odorless gas which is caused by incomplete combustion and it can be deadly. A carbon monoxide monitor should be installed near any of these units, and on each level of a building which receives forced air. Proper maintenance, cleaning and adjustment can help keep the systems running safely and efficiently. Higher efficiency furnaces produce higher levels of moisture which are disbursed at a lower heat level and typically vented through PVC. Although less drying, these more moist gases can also be corrosive. (See the "Environmental" section for more information on dangerous gases in the home).

HEAT PUMPS

A heat pump is a forced air system that provides both heating and cooling cycles. Most heat pumps are generated by electricity, however there are gas units available. The heat pump moves heat from one place to another. Heat is absorbed at the evaporator in one location, and released through the condenser in another. By reversing this process (reversing valve) the evaporator and condenser can exchange functions, thus enabling the system to heat and cool.

The compressor (outside unit) changes the refrigerant vapor from low pressure to high pressure. The cycle begins with the refrigerant in a liquid state under high pressure; the liquid is then forced through a pressure reducer and capillary tube. From there, it passes through the evaporator under low pressure. The liquid in the evaporator "boils" as it absorbs heat and is turned into vapor. The vapor flows into the compressor and the compressed vapor increases in temperature and is forced through the exhaust valve. The high pressure vapor passes through the condenser where heat is released into the surrounding air. Once the heat is released, the refrigerant turns back into a liquid, and the cycle begins again. The heat exchange occurs in the plenum as air passes over the coils (condenser when heating, evaporator when cooling) near the blower, and is delivered through the trunk to the supplies, thus heating the building. This system has limitations once the temperature drops below freezing outside.

These units are more efficient in the warmer southern states. However, their popularity is growing in the north. People often complain about the air not feeling hot when on standard heat mode. This is due to the limited amount of heat the unit can pull from the outside air. In the north and colder regions, heat pumps have to have a backup system. Electric elements (resistance heating) is one of the more common backup systems available; the coils are inside the air handler and engage as the unit demands, or when switched to emergency heat. Back up systems may also be oil fired, gas, solar, or other methods. These systems utilize air to air heat exchange.

Larger buildings may have more than one heat pump unit; sometimes one unit per level. This creates multi-zones which can be controlled individually. In smaller buildings or retrofits, a wall unit with a built-in air return is common. It is not advisable to repeatedly switch the unit off and on, as some units have a delay timer built in; others do not and damage to the compressor could result. The pressure in the compressor needs to drop before the unit begins to cycle again, or the motor will have unnecessary strain put upon it.

Problems to look for: HEAT PUMPS

- *Noisy unit* (severe vibration), possibly due to compressor failing.
- *Oil leak* at compressor, possibly due to compressor failing.
- *Lack of heat*, possibly due to compressor failing or refrigerant in need of charging. Check thermostat and backup system for proper operation.
- *"Delta T"* about 15-24 degrees between the supply and return; A lower reading may indicate the system needs to be charged and a leak could be present. A higher reading could mean a dirty filter or inadequate air return.
- *Dirty filter* which could restrict proper air flow and ability to heat.
- *Missing filter* this will cause dirt to pass freely through the system and house, and could also cause buildup on the blower, thus reducing efficiency.
- *Poor access for filter*, makes changing the filter difficult, and in some instances, dangerous.

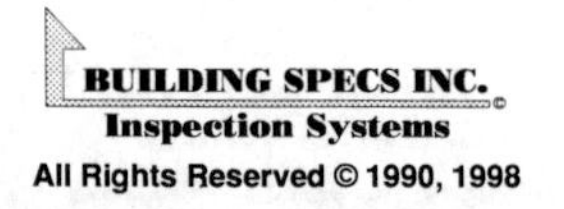

(exposed sharp metal edges with slide in cover, site made).

- *Missing overflow pan* under unit in the attic or second level units, could lead to water damage to the lower level ceilings.
- *Clogged condensation drain*, could lead to water flooding problem even under house.
- *Back up heat not working*; limited heat available in colder months. Test with clamp on amp meter.
- *Inadequate air return*, can over pressurize a room by not allowing air to return or escape. Can cause isolated cold rooms.
- *Rooms with no return*, doors tight to flooring. Closure of door reduces air into room, due to over pressurization.
- *Outside unit settling*, out of level, sinking in ground, resulting in rust. Can cause unit to prematurely fail.
- *Older unit* (poor condition), limited life span of 10-15 years average.
- *Ice* on evaporator coils.
- *Damaged or rusted evaporator coils.*
- *Water in the base of the unit*, clogged condensation drain.
- *Dirt or building debris* in the base of the unit.
- *Unit not running* when system is activated.
- *Missing or damaged insulation* on refrigerant lines.

 ***Maintenance may include:***

- Changing or cleaning filter.
- Correcting any of the preceding problems.
- Recharging system as needed.
- Annual service on units that have a combustion heat source.

GEOTHERMAL (Heat Pump)

Geothermal heat utilizes similar principles as air to air heat pumps. The outside coil utilizes the constant temperature below grade (depth varies in different regions) or in a body of water such as a lake. The lines run in the ground and may run horizontal or vertical. Some systems may have the temperature checked at self-sealing tubes, which allow for a thermometer to be used. At a minimum, feel the lines for varying temperatures.

OIL FURNACES

Oil fired furnaces are broken into two basic types, Forced Air and Hydronic (hot water boiler and steam). Any of the systems which involve a fuel that burns, such as oil, should be serviced annually. Any of these can produce carbon monoxide.

Proper maintenance, cleaning and adjustment can help to keep systems running safely and efficiently. The furnace should be installed in an area where it can get air (louvered doors or other).

The most common oil burner is known as a gun type. Oil cannot burn in a liquid state so it must be atomized (turned into a mist). The oil is is pressurized through a small nozzle, where it

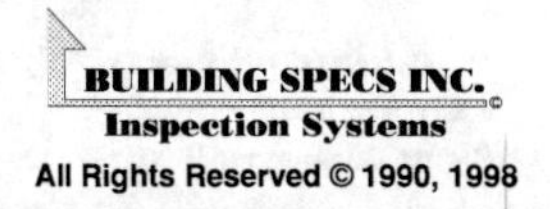

Oil fired furnaces require annual service. Nozzles and filters should be changed at each service.

is atomized and mixes with air. Once in the combustion chamber, an electric spark ignites the fuel mixture. Once the process begins, the ignition may be continuous, or just last until the initial fuel ignites. If the nozzle is poorly adjusted or becomes clogged, the unit may fail or burn inefficiently. The burner should run quietly and have a clean, even burning flame.

The burner fires into a combustion chamber where a heat shield is used to protect the unit. The combustion chamber must be leakproof, so that by-products from the combustion can not transfer into the air supply. When accessible via a view port, the combustion chamber should be checked for any cracks in the heat shield. Some replacement heat shields may be installed, but the design of the combustion chamber and heat exchangers may vary. Some retrofitted older coal-fired units may have the access sealed shut.

When the thermostat is raised, the furnace should start within a few seconds. If a severe time lag occurs there could be a problem. If there is a puff back due to a delayed ignition, fumes can enter the house.

 ***Problems to look for: OIL BURNERS***

- Noisy motor.
- Poor flame color.
- Excessive soot.
- Back puff.
- Flammable materials near by.
- Oil leakage.
- Pulsating sound.
- Excessive corrosion at view port.
- Cracks in heat shield.
- Improper flue connections, failing condition.
- Improper flue slope.
- Flue failing or corroded.
- Missing or non-functional damper.
- Susceptibility to supply line freezing.
- Visible oil leaks.
- Supply lines susceptible to freezing.

✔ ***Maintenance may include:***

- Annual service, cleaning and adjusting burner as needed, check and clean nozzle, replace filter cartridge, check all connections, clean strainer, adjust pressure setting.
- Repair or replace heat shield as needed.
- Inspect flue stack.
- Oil motor if applicable.

Gas forced air heat exchanger burners should be inspected for a steady blue flame and no change in flame when blower is activated.

GAS FURNACES

Gas burners come in several styles, though the concept is the same for most of them. The fuel (gas) and air pass through an orifice and are combined (venturi). The mixture flows to the burner where combustion takes place. The burner may consist of several components including the orifice, burner head, pilot valve, manifold, pressure regulator, manual and automatic shut off and air inlet. There are three basic burner manifolds: a series of holes, slotted, and straight, which is force-blown to a target. Gas furnaces can utilize air, water, or steam as a heat source.

Some gas furnaces utilize a blower to force the combustion flue gases outside. These should be checked for proper operation.

There are two basic systems used to ignite the gas. The first is a pilot light, a constant flame that is fueled by a supply from the pressure regulator. A pilot light is also equipped with a safety device, either a thermocouple or a thermal fluid bulb, which will shut down the gas supply if the pilot light goes out.

The second method used to ignite the fuel uses electronic ignition. Electronic ignition can consist of an electronic spark igniter or a glow coil.

There are three types of gas fuels used:

- Natural (comes from ground deposits)
- Manufactured (comes from distilling or cracking coal or oil.)
- Liquefied petroleum (LP/Gas) normally propane with a little butane added. Propane is considered more dangerous, since it is heavier than air and can collect in closed off spaces.

All of these gases have an odor added to help you detect the presence of a gas leak. If you suspect a leak, the gas should be turned off, and the windows should be opened. A qualified serviced technician should be called to locate and repair any leaks.

Most areas have specific codes about heating system installation, and many codes require gas furnaces to be installed in a ventilated area, not in an air tight area or a sleeping area.

 ***Problems to look for: GAS***

- Poor flame color.
- Flame rolling out toward the burner's opening (inadequate draft of the flue or vent).
- A change in the flame when the blower comes on (possible crack in the heat exchanger).
- Cracked or rusted-through heat exchanger.
- Excessive rust at the heat exchangers.
- Rust piles accumulating on the burners.
- Delayed ignition.
- Gas odor (service immediately).
- Missing drip/dirt leg.
- Corroded or rusted supply lines.
- Electrical ground to gas line.
- Noisy burner.
- Noisy blower.
- Corroded flue.
- Improper flue connections.
- Soot or fumes.
- Soot around the register areas.

 ***Maintenance may include:***

- Annual service to the unit is recommended.
- Clean and light pilot light.
- Test pilot safety device.
- Check and clean burner assembly.
- Check and clean any heat exchanger.
- Check and adjust flame.
- Check installation and condition of flue or vent.
- Follow maintenance that applies to forced air or hydronic systems.

FORCED AIR

This is a fairly efficient source of heat. Forced air may be heated by means of electrical resistance, heat pump, oil, gas, solar or a heat exchange with hydronic coils (where heated liquid passes through coils in the air handler where a heat exchange occurs). This heated liquid system is not as dry a heat as other forced air methods, and it is also safer, since the chances of carbon monoxide entering the supply air are diminished if not eliminated.

These systems use a blower to force warmed air into needed areas. As air is introduced into the room it becomes pressurized, forcing the existing air to circulate back to the system through the air return. A common problem, however, is when there is a central air return in a hallway. If the doors into the rooms are tight to the carpet, the room becomes over-pressurized. When this occurs, the air being supplied into the room diminishes, and can result in a cold room. One way

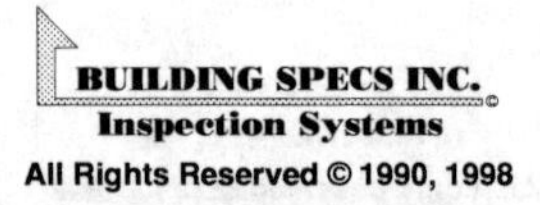

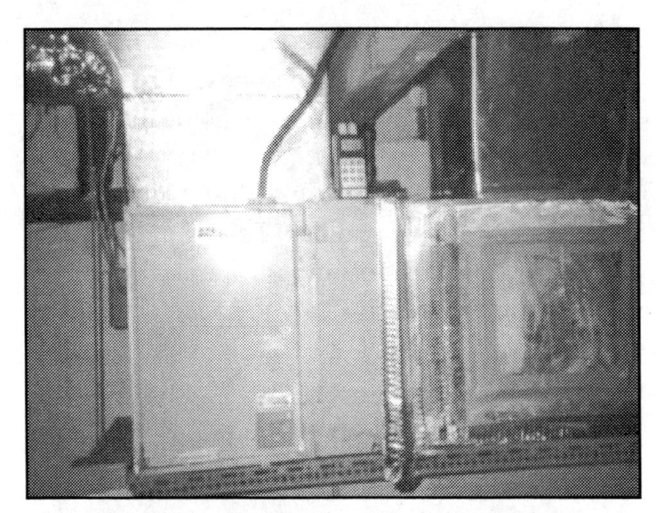

Heat pump air handlers. There should be a difference between the supply and return air of approximately 15 to 24°.

to test this is to measure the air velocity coming from the supply with the door open and then closed. If there is a significant difference, steps may have to be taken to correct the situation. These may include trimming the door or installing an additional return.

Forced air can be very dry, and a humidifier may be necessary to add moisture to the air. As air passes through the heat exchanger, it is heated by the rising hot air from the combustion chamber. The air is then circulated by the blower (commonly a squirrel cage) and delivered to the supplies and registers. The registers may act as a zone by closing or opening the damper control.

 ***Problems to look for: FORCED AIR***

- Dirty filter (restrict air circulation).
- Missing filter (allow dirt to pass through system and building freely).
- Dirty blower (inefficient air movement).
- Noisy blower (loose belt, needs oil, bad bearings).
- Older units (may need service).
- Excessive corrosion.

 ***Maintenance may include:***

- Changing filter.
- Clean blower (squirrel cage).
- Oil motor, tighten belts, if applicable.

HYDRONIC (Hot Water)

This title usually refers to a hot water circulating system. The water may circulate through finned baseboard, cast iron baseboard, concealed radiant pipes or radiators. Cast iron is an extremely efficient surface which absorbs and retains heat from the water, and radiates as well. Hydronic radiant heat, widely used in the 1960's, has made a come back recently. Since the circulation lines are buried in the floor, inspection is not possible. These are all very efficient, comfortable heat sources. They are also safe

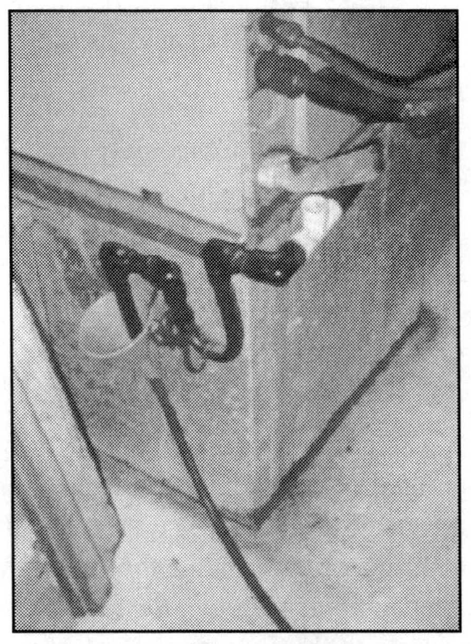

Condensate drain for air conditioning should terminate to outside air source. Above is an example of an improperly installed drain.

Water line in steam boiler should be at appropriate level in sight glass.

heat sources, since there is no ductwork for combustion fumes to pass into the house. Frequently, this heat has separate zones, so sections of the house can be controlled independently.

Some hydronic systems utilize a closed loop which runs through a forced air handler. This is a very safe form of heat since no carbon monoxide can enter the air system through the ducts. This is not as dry a heat as forced direct heated air, and is similar to a geothermal system. These systems are limited for visual access, and should be serviced yearly.

⚠ ***Problems to look for with hot water:***

- Visible leak (at boiler).
- Rust or corrosion (possible leak in system).
- Visible leak (at connections).
- Leaks at radiators, baseboard, circulator pumps (fittings or seals failing).
- Noisy circulator pumps (pump could be failing).
- Broken circulator pumps (zone not heating).
- Missing expansion tank (system could become over pressurized).
- Mixed metals supporting expansion tank from ceiling (metal straps could fail due to galvanic or other metallic reaction).
- Older units in poor condition (may need service).
- Noisy baseboards or radiators (air in the line).
- Water logged expansion tanks.

 ***Maintenance may include:***

- Repair any leaks
- Annual service
- Bleeding air from system

STEAM

Steam heat is an excellent form of heat. Steam is created in a boiler, and it rises to the higher parts of the heating circuit. The steam is generally at 212°F (100°C) or greater. As the steam is released into the needed areas (through the radiators) it condenses, and returns to the furnace boiler.

There are two types of systems used, single pipe and double pipe. The single pipe uses the same pipe to carry the steam and return the condensate. The two

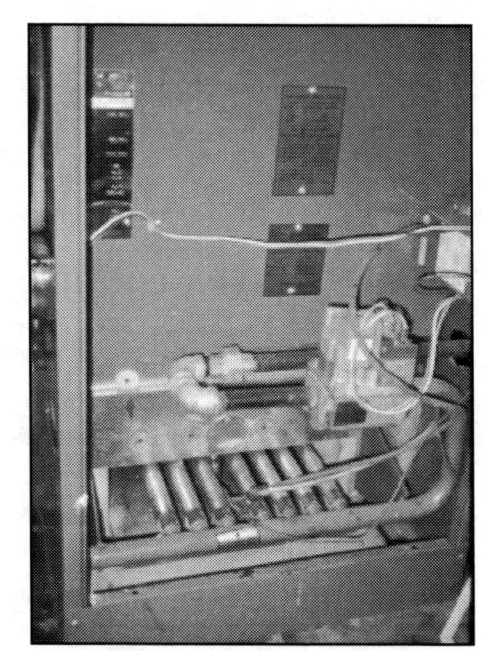

When inspecting gas or oil boiler look for signs of rust or leakage.

pipe system uses one pipe to supply the steam and one pipe to return the condensate.

The radiators create a heat exchange with air and do not release any humidity into the air. Heat is circulated by thermal movement.

The boilers need to be level and have the pipes sloped back toward the furnace. Air vents need to be mounted at the high points of the system. Each radiator should have a steam trap which allows only condensation to return to the boiler.

Check the water level in the boiler, it should be 1/3 to 1/2 full. If the boiler is empty, shut the system off at once. Do not introduce water into the boiler until the unit has completely cooled, or the boiler could crack.

A steam boiler must have:

• Pressure safety valves.
• Water level gauge.
• Pressure gauge.
• Temperature gauge.

 ***Problems to look for: STEAM***

• Low water level.
• Over pressurization.
• Visible leak (at boiler).
• Rust or corrosion.
• Visible leak (at connections).
• Leaks at radiators.
• Missing expansion tank.
• Older units.
• Leaky pressure relief.

Older furnaces may exceed their life expectancy but should be routinely monitored for proper operation.

 ***Maintenance may include:***

• Annual service.
• Repair any leaks.
• Maintain proper pressure.
• Maintenance on burner mentioned in other sections .

SOLAR HEAT

Solar energy is becoming more popular since it is a clean, non-exhaustive resource. Solar radiation may be utilized in two basic ways, passive and active. Passive heat would incorporate building orientation, window placement, floor absorbing material, or a greenhouse room to trap heat, and usually does not require blowers or mechanical components. An active system would include a collector, a circulator, and a storage system. Many of the systems are customized, combining existing systems for use as back up or primary, including:

• Hot water supply.
• Hot water heat.

• Heating forced air (heat pump).
• Generating electricity.

An active solar heating system should have the following components:

• *Solar collector*, used to trap the sun's radiation. Generally a black surface on an insulated surface with a glass cover which is used to trap the heat. Special surface coatings are being developed and used to increase the efficiency of the heat absorption. The cover protects the absorber, allows for the sunlight to transmit to the absorber, and traps the heat in the unit.

• *A circulator* which pumps a liquid, usually water with ethylene glycol added, to prevent freezing in colder climates.

• *Storage units* may consist of either an insulated holding tank for water, or a place where heat exchange can occur with air. An insulated rock storage container can be used to trap warm air. Generally, these containers contain larger rocks at the bottom, with smaller stones on top.

 ***Maintenance should include:***

• Keeping the glass surface clean and free of debris.
• Maintaining fluid levels and antifreeze.
• Oil circulator pumps or motors as needed.

ELECTRIC HEAT

Also referred to as resistance heating, electric heat works on the principle of natural air convection. Heavier, colder air sinks and the warmer, heated air rises, passing over the heating elements. Some units utilize an oil or liquid filled element, which makes the heat feel more like hot water baseboard heat, and not as dry as conventional electric baseboard heat. Electric furnaces are simply an air handler with resistance coils. These utilize a forced air system.

Some of these units have thermostats located directly on the units, while others use a wall mount thermostat. Most of these units utilize a 220 V/20 amp breaker per circuit. The units may vary in size, ranging from twenty four inches to several feet.

 ***Potential hazards***

Many of these units will achieve temperatures in excess of 200° F and could cause:
• A severe burn or scalding to human contact *(children).*
• Fire with contact of drapes, furniture, etc.
• Shock or fire when electrical cord comes in contact with the unit.

 ***Maintenance may include:***

• Keeping the units clean and free from debris. There should not be any electrical outlets or drapes over or touching these units, as this could present a fire hazard.

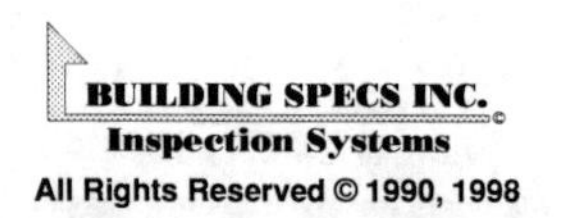

Humidifiers can cause excessive rust and corrosion to duct work and heat exchangers. Unmaintained humidifiers may allow bacteria to grow and circulate into air supply.

HUMIDIFIERS

With a forced air heating system, the air in the house may become very dry. This can be uncomfortable to the occupants in the house, and can cause dry skin and breathing difficulty. The relative humidity should be around 35%. Low humidity can also dry out woodwork, furniture, and other possessions in the house. Humidifiers add water vapor to the air and can be introduced directly into the heating ductwork, or be an entirely independent system. Excessive humidity may cause mildew and molds to form, so it is sometimes beneficial to have humidity levels controlled by a humidistat.

There are several styles of humidifiers:

- Drum
- Mist
- Evaporation
- Electric

 ***Maintenance may include:***

- Annual service and cleaning. Hard water or well water may make frequent cleaning necessary. The nozzles and drains are susceptible to clogging, and floats should be checked for sticking. If the reservoir is allowed to overflow, the water could back into the furnace and cause damage and corrosion. Some humidistats are difficult to service and may require a heating specialist. If there is a lot of corrosion on the unit, it should be serviced.

A blocked flue damper could cause hazardous backdrafting.

DUCTWORK

Ductwork refers to the material that supplies air to the rooms, and returns it back to the unit. The furnace warms the air and feeds it to the main supply trunk. Smaller branch supplies are run off of the main trunk, and are terminated at walls, floors, or ceilings by means of air registers or diffusers. The air then returns back to the unit and the cycle is repeated. Different systems may have the blower and filter placed before or after the heat exchange. Some systems may utilize two sets of air returns, alternating closing, depending on whether the unit is on heating or cooling cycle.

There are several materials used for ductwork. They include:
- Ridged metal (galvanized).
- Ridged metal (galvanized/insulated).
- Ridged insulated.
- Flexible(uninsulated).
- Flexible (insulated).

 ***Problems to look for: DUCTWORK***

- If the ductwork is exposed in an attic or other area affected by temperature, it should be insulated.
- The ductwork should be well supported and connected to the trunk line securely.
- Flexible duct work should have gradual smooth bends and never be kinked.
- A supply should never be run directly off the end of a trunk, for this will increase the air to this supply and choke off the previous supplies.
- Look for any signs of corrosion or water damage, which could indicate problem.

FLUE STACK

There are different tests to measure flue stack gases and temperature; these are usually carried out by a qualified service technician. These tests can help determine the unit's burning efficiency.

Inspect the condition of the flue, look for corrosion or deterioration. Some corrosion might come from water running down from the exhaust, due to the lack of a flue cap. A condensation trap is recommended to prevent moisture from backing into the furnace.

The flue should slope up and away from the furnace, to help create a draft, and pull the gas up through the flue. There should be a damper control on the flue to help with the air draft. While inspecting the flue connections, you can compare combustion fumes in the flue with running water. If the water in the flue was running away from the furnace, could it leak out at any of these connections? The connections must be tight, must consist of male to female fittings, and lead away from the furnace. If not, the combustion gas flow could leak out at the connections into the house and could cause asphyxiation.

Some gas furnaces may be fitted with a blower, which should come on prior to ignition, to insure that the fumes are expelled.

THERMOSTATS

The thermostat is used to control the heating or cooling system separately or in a combined unit. The thermostat should be located in a central area of the house. It should not be located near a heat source, window, exterior door, or behind a concealed or covered area. Any of these locations could influence the thermostat's reading of the room temperature result in an inaccurate reading, thus causing improper heating or cooling of the house. There are several types of thermostat connections which react to temperature change:

- Bimetal strip

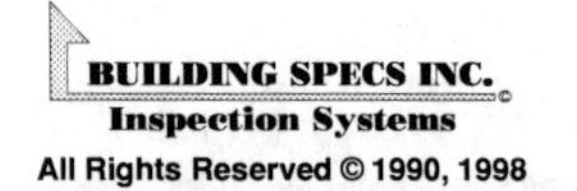

- Rod and tube
- Bellows or diaphragm
- Electrical resistance
- Hydraulic

Thermostats will either be a line voltage thermostat (such as electric baseboard) or low voltage which is connected to a relay switch. The low voltage gets its power from a step-down transformer. Some thermostats come equipped with heat anticipators, which prevent the heat from shooting past the desired temperature.

Mercury switches are very common in residential buildings. They are considered a bimetal switch. Check to see if the mercury flows freely and has not dried or become sluggish.

Cooling

AIR CONDITIONING SYSTEMS

Air conditioning units work under the same principle discussed in the section on heat pumps. Heat is collected at the evaporator inside, disbursed through the condenser and distributed to the air outside. Air conditioners may be shared (split) with a forced air furnace, a separate system, or a wall or window unit.

The most common cooling systems are electric, however there are gas-fired units as well. Some systems utilize ammonia and even salt water for refrigerant instead of freon. The air conditioner should not be tested if the outside temperature has been below 60 degrees for 12 hours.

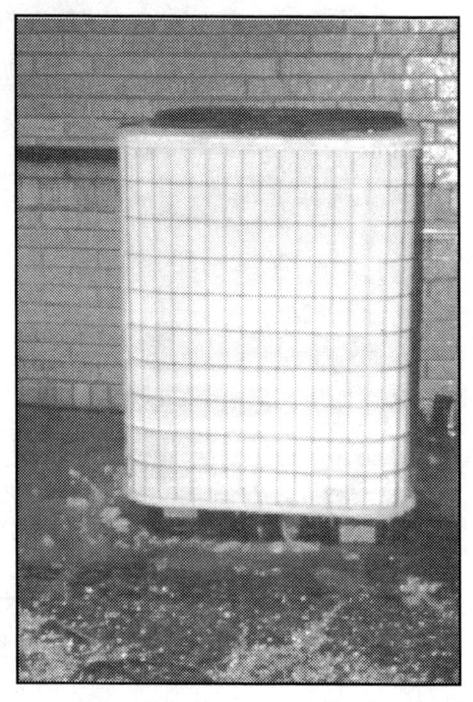

Heat pump compressors icing indicate a problem with the system.

 ***Maintenance should include:***

- In winter, the units should be covered to protect them from ice and possible damage.
- See heat pump maintenance.
- Change or clean filters.

FILTRATION

Indoor air pollution is a growing concern.

Air contaminates include:

- Solids (dust, fumes, smoke)
- Liquids (mist)
- Gases (carbon monoxide, hydrocarbons)

The job of the filter is to trap some of these contaminates and other airborne particles, prevent them from passing through the system (damaging the unit), collecting on the blower, and being recirculated into the air. The average residential filter traps only the solids.

There are several types of filters available:

Disposable filters

- Adhesive filters are very common. The fibers are usually coated with an adhesive to help trap dust.
- Disposable filters come in many styles and are made from a variety of materials, including, glass, cotton, and synthetics.

 ***Maintenance should include:***

- Replacing the filters as they become clogged, usually a minimum of twice a year. If there are pets or plants in the house, the frequency may increase. Hold a bright light up to the filter to look for heavily loaded areas, or any holes in the filter itself.

Washable or Reusable Filters

- Washable filters included aluminum and synthetics (many different styles available). Stay away from frameless filters, they are susceptible to failure, and can be sucked into the plenum, allowing dirt to bypass. Some washable filters are adhesive as well. A spray can be utilized to enhance the adhesive coating.
- Pleated or deep pocket type filters, allow for a larger filtering surface. Some are reusable and others are disposable.
- Electrostatic filters use a passive charge to attract and hold the particles. Some may restrict air flow and cause a pressure drop across the coils.
- Electronic air filters: These filters are a very effective method of cleaning the air. As air passes to the filter it usually passes through a pre-stage filter that traps larger particles. The air then moves to the electronic filter where it passes through a highly ionized field. A series of wires puts a high positive voltage on the particles as they pass through the field. The particles are then attracted to the negative grounded plates.

 ***Maintenance should include:***

- Cleaning every few months. Most washable filters can be put in a dishwasher or hosed off. Check to see what the manufacturer recommends.

These filters however, are not always successful in removing odors and other gases.

There are a number of specialty filters available which include:

- Carbon (removes odor and bacteria).
- Ultraviolet light (ozone producing, which kills bacteria).

HEATING & COOLING CHECKLIST

Heating System

☐ *Service Recommended* ☐ *Fuel Leak: Tank / Supply Line (Service Immediately)* ☐ *No Sign Of Recent Service*

1] **TYPE OF SYSTEM** ***[See Lower Section For Additional Information On Heat Pump(s)]***

☐ **FORCED AIR:** ☐ **Heat Pump** *(Air to Air)* ☐ **Heat Pump** *(Geothermal)* ☐ **Electric** ☐ **Fossil Fuel** ☐ **Hydronic** *(Coils in air handler)*

☐ **HYDRONIC:** ☐ **Radiant;** *(Floor, Ceiling)* ☐ **Baseboard** *(Fin, Cast Iron)* ☐ **Radiator** ☐ **Steam** ☐ **Solar** *(Not part of standard inspection)*

☐ **ELECTRIC** ☐ **Baseboard** ☐ **Wall Unit** *(With Blower)* ☐ **Electric Radiant** *(Floor, Ceiling)* ☐ **Portable** *(Potential Fire hazard)*

☐ **CENTRALIZED SYSTEM** *(Heat not directly supplied to all rooms)* ☐ **SEPARATE UNITS** ☐ **Other**________________

2] **FUEL SUPPLY** ☐ **Electric** ☐ **Oil** ☐ **Natural Gas** ☐ **Propane** ☐ **Solar** ☐ **Wood, Coal**

3] **CONDITION** ☐ Ok ☐ Newer Unit MFD________ MFD________ ☐ ***Not Operational*** *(Have Serviced, and Certified By HVAC Contractor)*

☐ **Problem;** Inadequate or Limited Heat Supply / Restricted, No Access / Motor Noisy / ***Amateur, Sloppy Installation / Unsafe Installation / Older***

☐ ***Excessive Rust or Corrosion Noted At;*** Plenum / Heat Exchanger / Observation Port / Duct / Flue / Unit Casing / Near Pressure Relief Valve

4] **BURNER** *(With fossil fuel system)* ☐ Ok ☐ **Problem;** Possible Cracks In Fire Box / Possible Cracks In Heat Exchanger / Poor Flame Color Burner Not Firing / Poor Flame Shape / Flame Rolling Out / Inadequate or Restricted Air Supply To Unit ***No Fuel / Fuel Turned Off*** *(Not tested)*

5] **FILTER** ☐ Yes ☐ **Disposable** ☐ **Reusable** ☐ **Electronic** ☐ ***Located In The;*** Air Handler / Air Return / Ductwork

☐ **Problem;** Missing / Poor Access / Missing Cover / Dirty / Failing / Sucked Into Unit / Non Functional / Too Small For Opening / Not Located

6] **AIR HANDLER** ☐ Ok ☐ **Problem;** Coils Dirty / Blower Dirty / Coils Icy / Coils Icing, Damaged / Standing Water In Unit / Unit Noisy / Rusted

7] **DUCTWORK** ☐ **Rigid Metal** ☐ **Flex Metal** ☐ **Plastic Flex** ☐ **Insulated Fiberboard** *(Fiberglass lined)* ☐ **Ductless** *(Plastic Tubing)*

☐ **Problem;** Inadequately Supported / Restricted Air Flow / Possible Asbestos / Air Return Inadequate / Damaged / **Dirty;** Ductwork, Air Return Run Off End of Trunk / Wet / Inadequate Connection(s) / Disconnected Connections / Insulation Deteriorating / Sloppy, Amateur Workmanship

8] **FLUE** ☐ Ok ☐ **Rigid Metal** ☐ **Flex Metal** ☐ **Plastic** ☐ **Masonry** ☐ **Other**________________

☐ **Problem;** Corroded / Deteriorated / Unable To Fully Inspect / Excessive Soot / Reversed Connection / Inadequate Connection / ***Hazardous*** Poor Condition / Inadequate Slope / Possible Asbestos / Damper Missing / Damper Stuck, Cemented Shut / Sloppy, Amateur Workmanship

9] **THERMOSTAT** ☐ Ok ☐ **Problem;** Damaged / Not Operational / Inadequate Installation / Not Operating Properly / Poor Location

•ADDITIONAL COMMENTS *•Heat Systems are checked for operation and no remaining life expectancy or warranty is implied. [1]An older unit may be nearing or at the end of life. Heat exchangers should be inspected and certified by a heating contractor. Systems require routine maintenance. MFD abbreviated for manufactured date and may not be readable on unit.*

____

____

Heat Pump / Air Conditioner

☐ ***(A.C. Unit Not Checked in Temperatures Which Have Been Below 60°F in the Past 24 Hours.)***

1] **TYPE OF SYSTEM** ☐ **Heat Pump** ☐ **Air Conditioner** ☐ **Shared** ☐ **Window/Wall Unit** *(Not part of Inspection)* ☐ **Gas Unit**

2] **CONDITION** ☐ Ok ☐ Newer Unit MFD________ MFD________ ☐ ***Not Operational*** *(Have Serviced, and Certified By HVAC Contractor)*

☐ **Problem;** Restricted Access / Noisy / **Overflow Pan;** Missing, Standing Water / *Sloppy, Amateur, Inadequate Installation* / **Older Unit**[1]

3] **COMPRESSOR UNIT** ☐ Ok ☐ **Problem;** Not Level, Elevated / Noisy / Damaged / Rusted / Coils Damaged / ***Coils Appear To Be Icing*** Unit Ceased / Unit Not Operational / Insulation On Refrigerant Lines Missing, Deteriorated / Oil Leak Observed / Fan Vibrating / **Older Unit**

a] ☐ **Refrigerant Lines Temperature** ☐ Variation Noted In Lines *(Good)* ☐ **Problem;** Close / Same *(May be a sign of low refrigerant, or compressor failure)*

#1 Temperature Reading Supply______° Return______° ☐ **Inadequate** *(May indicate low refrigerant)* ☐ **Too High** *(May indicate inadequate return)*

a] ☐ **RLA** *(Running Load Amps)* **[Recommended______Amps / Observed______Amps** *(If RLA observed meets or exceeds recommended compressor may be failing)*

#2 Temperature Reading Supply______° Return______° ☐ **Inadequate** *(May indicate low refrigerant)* ☐ **Too High** *(May indicate inadequate return)*

a] ☐ **RLA** *(Running Load Amps)* **[Recommended______Amps / Observed______Amps** *(If RLA observed meets or exceeds recommended compressor may be failing)*

4] **CONDENSATE** ☐ Ok ☐ **Problem;** Drain Clogged, Leaks / Inadequate Installation, Termination / Could Not Operate Pump / Pump Failing

•ADDITIONAL COMMENTS ***•Air conditioners and heat pumps have a life expectancy of 8 to 12 years.*** *Air Conditioners and Heat Pumps are checked for operation and no remaining life expectancy or warranty is implied. [1]An older unit may be nearing or at the end of it's life. See additional information for heating systems, ductwork, filters, etc., above. Running Load Amp (RLA) test is optional.*

____

____

____

BUILDING SPECS INC.©
Inspection Systems

Chimneys & Burning Systems

For any chimney unit, whether masonry, insert, or free standing woodstove, proper installation and maintenance are essential for safe operation. The consequences for a poorly installed unit can range from inadequate function, to fire hazard and smoke or carbon monoxide inhalation, to the death of the home's occupants. Any unit that burns a fuel produces combustible gases which can be harmful when inhaled. Carbon monoxide can be deadly, and is colorless and odorless. A poor flame (due to lack of air) can cause higher amounts of carbon monoxide; a proper flue system will carry these gases out to a safe point of exit. It is a wise idea to have carbon monoxide detectors placed in areas where these units are used. There are many situations and codes which will dictate proper and safe installation: a local fire inspector, housing inspector, or woodstove shop should be consulted.

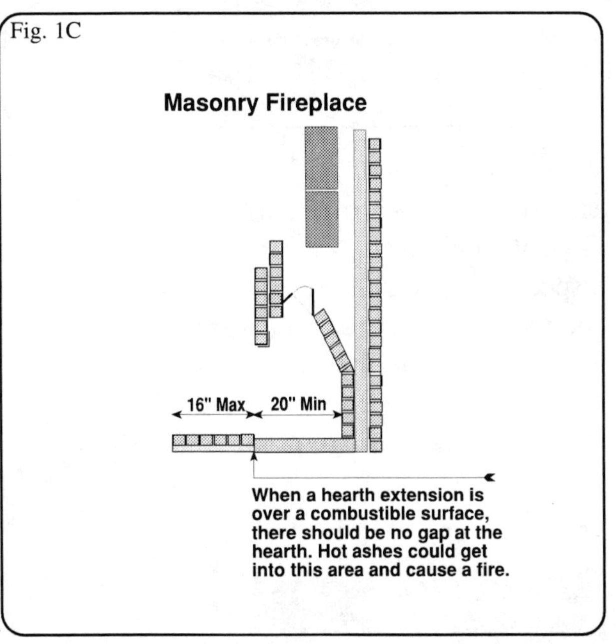

Deteriorating mortar may indicate water infiltration and unseen damage.

There are a few basic components of a wood/coal heating system.

1. Firebox (where combustion takes place)
2. Damper (to regulate draft and close the unit off)
3. Flue (to induce draft through air convection and carry the flue gases to a safe point of exit. (See Fig 1C)

These components will vary due to manufacturers' preferences, local codes, and design. All require annual inspection

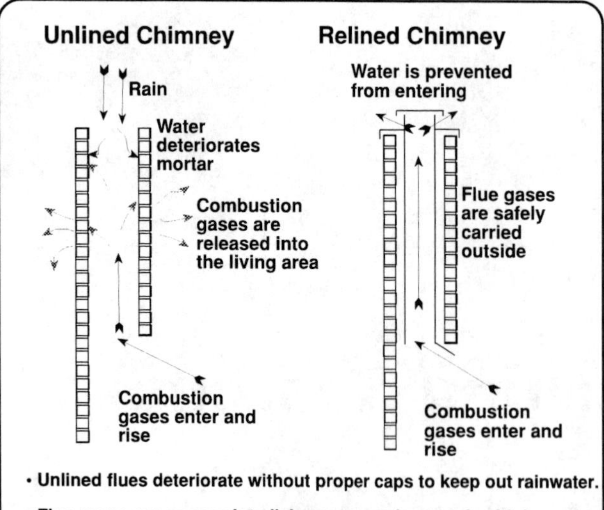

Fig. 2C

and maintenance for proper performance and safety.

CHIMNEYS

The chimney can encase a flue, which can be metal or masonry. Sometimes, typically in older buildings, the chimney itself can act as the flue, (See Fig. 2C), but an unlined flue can become a health and fire hazard.

Deteriorating mortar can allow flue gases and smoke to penetrate the chimney and enter the living space, and these gases can be deadly. A deteriorated unlined chimney can also be a fire hazard if the mortar or brick has deteriorated enough that the heat or flame can ignite a combustible surface.

If your chimney is unlined, a thorough inspection is highly recommended. The chimney can be pressurized and monitored for air loss, but the safest course is to simply reline the chimney. A continuous stainless shaft can be inserted, followed by a proper cement or metal cap. There are professionals who specialized in these procedures.

Unlined chimney flues may allow combustion gas to escape and water to infiltrate and damage systems.

A stainless steel flue liner makes for a proper upgrade to an unlined chimney.

The chimney can be constructed from masonry, wood, or the flue stack alone. If the siding is excessively cut away from around the chimney, or goes behind the chimney, it is a good possibility the chimney was a later addition to a completed home.

Masonry Chimneys

If you have a chimney made of masonry, inspect the condition of the mortar. Any decorative steps in the brick may expose the holes in the bricks. These areas allow water to penetrate and can cause water damage, and consequently should be filled with a hydraulic cement. If the chimney is on an outside wall, see if there is an excessive gap at the house, an uneven gap, or a cracked caulk joint at the siding. Any one of these could be due to normal settling or a symptom of the chimney shifting. If the top of the chimney is accessible and the chimney is in good condition, grab the flue and *gently* try moving the chimney away from the house. If there is excessive play further evaluation is in order. If the chimney is unlined, there are loose bricks at the top, or if there are any visible cracks or deterioration, extreme caution and discretion should be used: the top of the chimney could fall in this situation. If there is deterioration or other problems, a qualified specialist should be consulted.

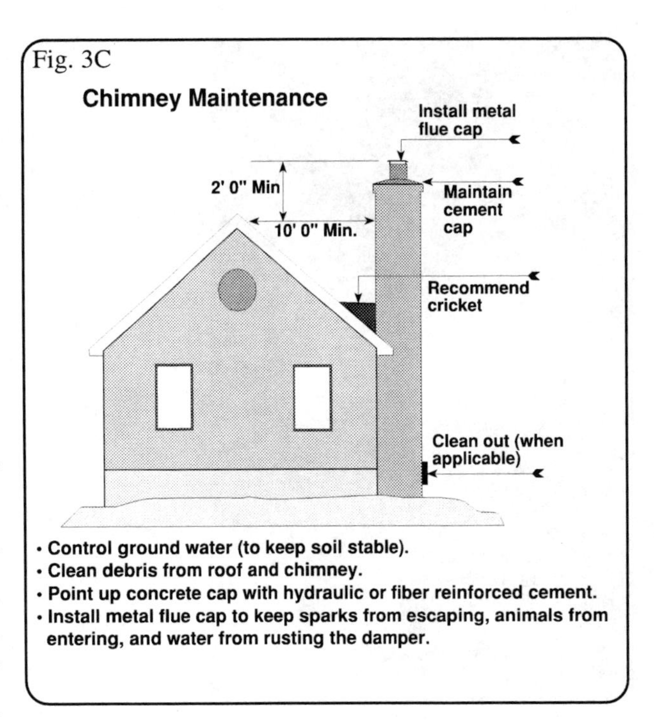

Excessive or perpetual settling could be due to:

• Inadequate footing: Dig down and verify depth and width. The chimney may need to have an underpinned footing added. Consult an engineer.

• Inadequate grading and saturated soil: Ground that is subjected to standing water or constant saturation may become unstable. The soil's weight bearing capacity can change. Try proper water control and ground cover. Allow the soil to stabilize and dry.

• Inadequate connection to building: There are through bolts available with star anchors --- decorative metal exposed reinforcing collars --- which must be properly installed.

Some movement in a chimney is normal. However, monitor the chimney for any signs of movement. Other signs of settling may be bricks cracking, mortar joints failing, caulking joints stretching along the brick and siding, flashing at the roof pulling down into the roof or cracking in the wall covering (i.e. drywall). In severe cases, the chimney may have to be demolished, but usually the settling will occur and stabilize itself and may just require some point up. Some chimneys are built so that they run on an angle a long distance, and may be susceptible to failure due to the weight being past the footing and its center. In any case, an engineer or chimney specialist should be consulted to

To prevent water damage to chimney, a proper cement crown and flue extensions should be installed.

possibly save the structure.

Wood Chimneys

Wood chimneys or chases enclose a metal flue, typically from a zero clearance unit. If there is ground contact, look for signs of rotting. When a wood chimney is on top of the roof, inspect it for proper attachment to the roof. Sometimes they are merely nailed to the sheathing and sagging in between the rafters. Try gently moving the chase. If there is excessive movement at the sheathing, added blocking and fastening is recommended. In a high wind area, a wood chimney improperly built could blow over.

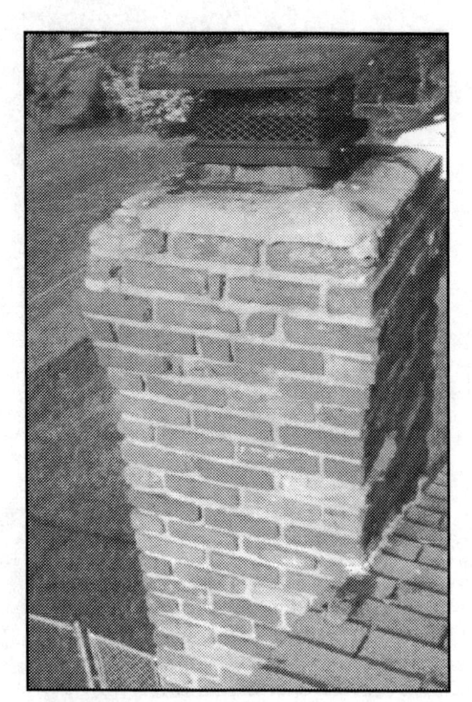

Spalling is caused by salt or water infiltration which results in brickface cracking.

CAPS

Any chimney should have a cap to prevent water damage to the unit that it is venting, as well as protect the integrity of the chimney. Metal flue caps (spark arrestors) should be installed as well to prevent animal or water entry, and prevent sparks from escaping and causing a fire.

Concrete Caps (Crowns)

Concrete caps fail at a significant rate, even on newer homes. The cap is subjected to extreme conditions, and when water begins to penetrate and freezes, the concrete starts to fail and crack. Once water gets between the brick and flue, water pressure creates a condition known as efflorescence, which is often responsible for the deterioration of the mortar joints. At this stage the cap needs to be pointed up to prevent any further damage. To prevent this from occurring, we recommend using a different cement for this application. Hydraulic or fiber reinforced cements are available, but they are costly. There is a waterproof portland cement that is available and should greatly increase the life of your cap and chimney. Inspect the bricks around the top few feet of the chimney for failing mortar, spalling (face cracked off from freezing), and loose bricks. (See Fig. 3C).

Metal Chimney Caps

Metal chimney caps are prone to sagging and ponding water which can lead to rusting. Any rust should be stopped and the metal should be sealed. With some skill, even the sag can be eliminated. The top of the cap should be screwed snug to the chimney, and note that if this top section is wood trim, it is prone to rotting.

CRICKETS

The cricket is essentially a small roof behind the chimney that deflects the water away and does not allow snow or leaves to pocket behind the chimney. Any chimney which has the roof falling into the back should have a cricket. CABO states a chimney over 30" wide and parallel to the ridge, should have a cricket. Rot and ice damming can occur on chimneys smaller than this. A cricket can prevent water and ice damage from occurring. This is even more critical when the chimney is beside a second level of an adjoining roof, because then water and snow can build up along the wall and work its way under the siding, causing the siding and possibly the frame behind it to rot. These areas should be routinely inspected and any debris should be cleared. New houses do not always have crickets installed. Crickets can be made from aluminum, copper, roofing material, rubber, and other materials, but whatever the material, flashing and counter flashing should be installed.

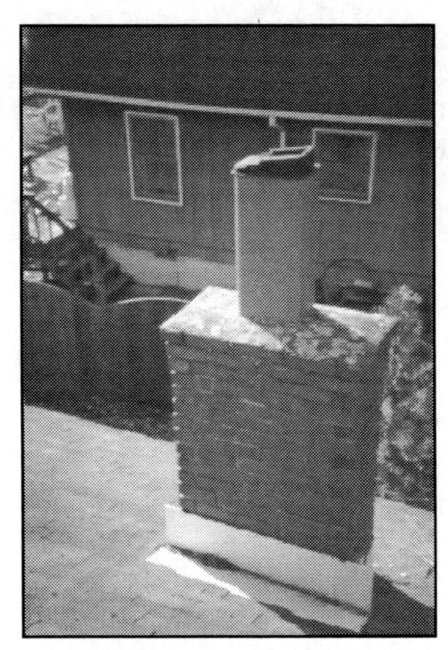

A retro fit damper on top of the flue replaces failing damper at chimney throat.

DAMPERS

The damper on a masonry, insert, or free standing woodstove, regulates the draft and shuts off the unit from outside air. Check for proper operation; it should operate freely and open and close securely. Commonly, a damper is broken or rusted out due to the lack of a flue cap. Broken dampers can be repaired in masonry units; some need new hinge pins while other need to be replaced. When reaching up and operating the damper be aware of animals (dead or alive) or nests that might be on the other side. Have something to wipe your arm off in the event soot falls.

FLUE

Flues are the passage in which combustion gases are carried outside by means of convection or a blower. The flue can be constructed from metal or masonry (fire clay flue liners, that resist temperatures up to 1800°F). Either of these flues should be routinely inspected and cleaned, for excessive soot could cause a chimney fire, and a masonry flue is susceptible to cracking and flaking due to water damage or an overheated fire. A common misconception is that it is good to stoke a fire and get it so hot that the soot burns out of the chimney. This can actually lead to the flue cracking.

There should be no gaps between the flues; the mortar joints should be tight. The top of the flue should stick out 3-4" above the cement cap, and a flue cap should be installed. Sometimes a flue will stick out of an unlined chimney to seal off the top, but this is not acceptable if the chimney leaks. If the flue is close to, even, or below the cement top, an extension should be installed.

Metal flues come in single and triple walls, depending on location. Proper, tight connections are essential, but limited for inspection. When a single-walled flue passes through a ceiling a heat shield is necessary, and these types of flues must be connected male to female, to insure the flue gas cannot escape into the living space. If excessive corrosion is present, the flue may need to be replaced. Look for water stains, which could indicate water entry at the cap or roof, and repair the situation should you find it.

CHIMNEY, FIREPLACE & FLUE CHECKLIST

1] **TYPE OF UNIT(S)** ☐**Fireplace** ☐**Insert** ☐**Free Standing Woodstove** ☐**Zero Clearance** ☐**Gas Unit** ☐**Heating System**

2] **FIRE BOX** ☐**Masonry** ☐**Metal** ☐**Brick Lined** ☐**Panels** ☐**Relined** ☐**Incomplete** ☐**Obstructed**

☐**Problem;** Needs Cleaning / Mortar; Deteriorating, Failing / Bricks Deteriorating / **[Cracking;** Minor / Severe] / Panels Cracking / Blocked Heavy Creosote / Burn Marks / Water Stains / Rust / Holes / Brick Face; Loose, Damaged / Burn Marks on Face / Mantel Too Close

3] **BLOWER** ☐Yes ☐No ☐**Problem;** Could Not Operate / Noisy / Amateur Installation / Sloppy, Hazardous Wiring / Not Connected

4] **DAMPER** ☐Yes ☐No ☐**Problem;** Damaged / Missing / Stuck / Loose / Falling Out / Rusted / Not Serviceable / Not Accessible

5] **FLUE**[1] ☐**Masonry** ☐**Metal** ☐**Unlined** ☐**Relined** ☐***Not Accessible For Inspection***

☐**Problem;** Heavy Creosote / Visible Cracking / Possible Relining Required / Missing Section(s) / Water Damage / Spalling At Sections Deteriorated / Gaps Between Tiles / Stops Below Cap / Rusted / Corroded / *Amateur, Sloppy Workmanship* /***Hazardous Condition***

6] **HEARTH EXTENSION** ☐Ok ☐**Brick** ☐**Slate** ☐**Marble** ☐**Tile** ☐**Concrete** ☐**Metal** ☐**None** *(Potential Fire Hazard)*

☐**Problem;** Missing / Cracked / Sloping Away From Unit / Gap at Fireplace Connection / Inadequate / Loose /***Potential Fire Hazard***

7] **CHIMNEY** ☐Ok ☐**Brick** ☐**Block** ☐**Stone** ☐**Wood** *(Flue chase)* ☐**Metal** ☐**Other**________________

☐**Problem;** Top Section Deteriorating / Mortar Deteriorating / Bricks Deteriorating / Spalling / Leaning / Loose Brick(s) / Mortar Cracking Rocking at Grade / Rocking at Loose Section / Rocking at Roofline / Separating from House / Hazardous / *Amateur, Sloppy Workmanship*

a] **Flashing** ☐Yes ☐No ☐**Problem;** Possible Leak / Susceptible to Leak / Missing / Need Cricket / *Amateur, Sloppy Installation*

8] **CAP** ☐Ok ☐**Cement** ☐**Metal** ☐**Brick** ☐**Other** ☐***Recommend Metal Flue Cap(s)*** *(To prevent water or animal infiltration)*

☐**Problem;** ☐*Not Accessible (Recommend further evaluation)* / Cracking / Deteriorated / Missing / Rusting / Sagging / *Amateur, Sloppy Installation*

☐**Cap (Crown) Needs To Be Upgraded Or Repaired** ☐**Recommend Flue To Be Extended Above Cap** *(To prevent water infiltration)*

•**ADDITIONAL COMMENTS** *[1]Internal components of the chimney are not covered under this inspection, such as the inside of the flue. A chimney specialist should be consulted for this. Annual maintenance must be maintained for maximum safety and to help sustain the life of the unit and it's components. Woodstoves should be inspected by a woodstove expert.*

Living Room: ____

____

Den: ____

____

Dining Room: ____

____

Basement: ____

____

Master Bedroom: ____

____

____

____

____

____

____

____

____

____

____

____

____

☐***Repairs Required*** ☐***Requires Further Evaluation*** ☐***Limited / No Access To Unit***

Section Six:

Electric Systems

Electrical Overview

Much of the electrical system is concealed, so inspection is usually limited. The opening of junction boxes, outlets, switches or other boxes is not part of a normal home inspection. A random sample of one or two boxes may be opened, though this still does not reveal all of the connections or any concealed splicing. Some older wiring connections may need updating, but this entails a separate thorough electrical evaluation.

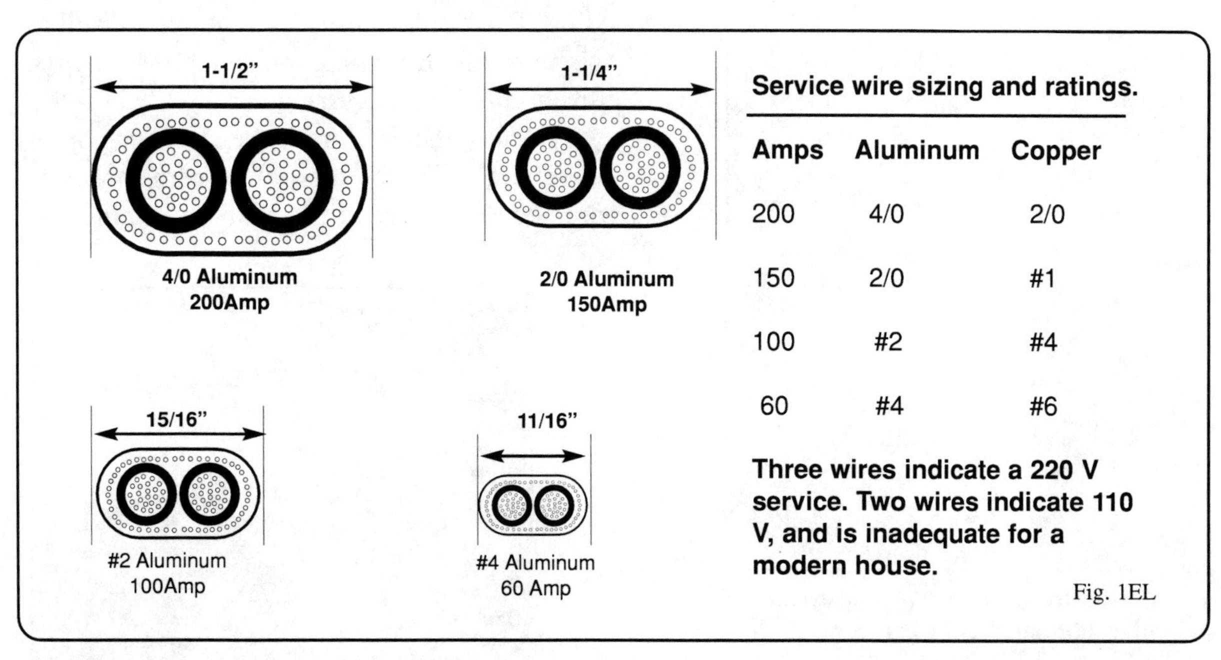

Service wire sizing and ratings.

Amps	Aluminum	Copper
200	4/0	2/0
150	2/0	#1
100	#2	#4
60	#4	#6

Three wires indicate a 220 V service. Two wires indicate 110 V, and is inadequate for a modern house.

Fig. 1EL

Age needs to be considered when inspecting the electrical components of your home. Older homes will typically have fewer receptacles per room than a newer home. Many historic homes have as few as one receptacle per room, or even none. Putting a multi-plug receptacle at one of these locations is not recommended. By introducing a large demand on the circuit, the frequency of blowing a breaker might increase. If a larger fuse or breaker has been installed, the fuse might not blow when needed, overheating of the wire can take place, and the potential for a fire hazard is present. Building codes have changed throughout the years, and houses that were built, for example, in the 1950's, would have to go through extensive renovation to bring the house up to current building codes.

It is impossible to predict if any given circuit will blow. Every family brings differing

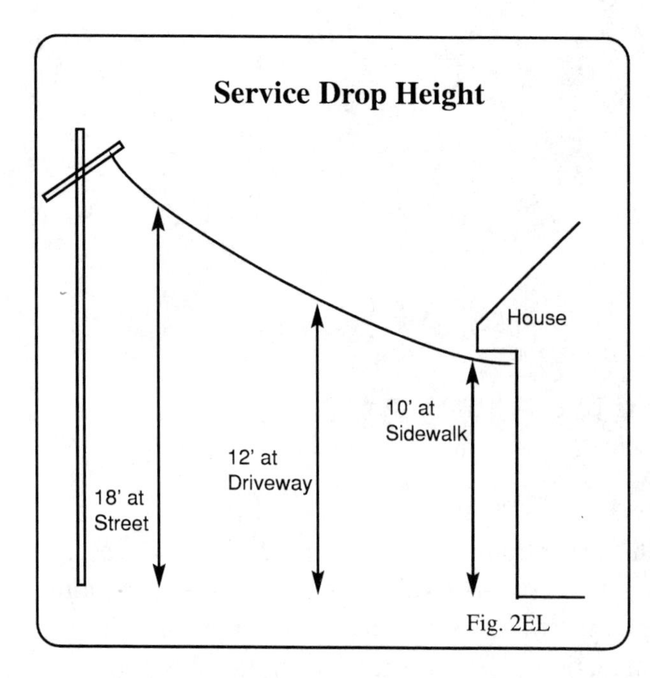

demands and loads to the same circuitry. One family could live in a house for several years with no problems, and then another family could bring several hairdriers, VCR's, a microwave, and kitchen appliances, and experience fuses blowing and over-loaded circuits.

SERVICE

There are some basic inspections that can be made. Any exposed exterior wire should be routinely checked for deterioration or fraying. Once the coating (sheathing) has deteriorated to the point the wires are exposed, the possibility of electrocution is present. When metal siding is in contact with the wiring, there is risk of the electricity conducting onto the siding. This could cause electrocution as well as a fire hazard. Grounding the siding may help to prevent this from occurring.

Typically the feed from the street or pole is the responsibility of the power company. Any wire section overhead needs to be a minimum (varying on region) of 8' from a roof and 12' from the ground. It is usually the power companies who keep the power lines clear from branches and handle other maintenance problems.

Two wire service drop indicates an antiquated 110 volts service.

The service wire dictates the amount of electricity the building can be supplied, not necessarily the breaker's size. For example, when a service panel has been upgraded from

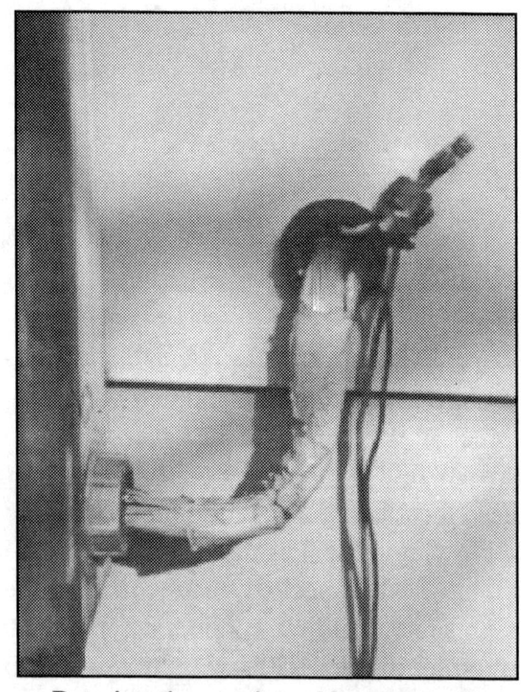

Deteriorating service cable may expose conductors and pose electrocution hazard.

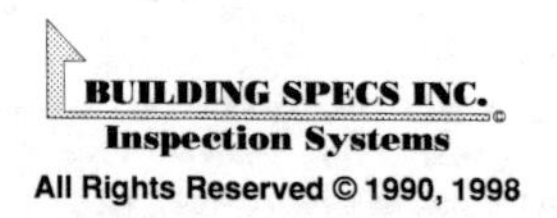

100 amps to a 200-amp main breaker, the service wire must be upgraded as well. When the wire is not upgraded, the panel is still only rated a 100-amp panel. Often the service wire and panel are upgraded, but the circuits are still outdated. For example, a single circuit running through a kitchen, which carries all the lighting and receptacles, as well as the microwave and refrigerator, is considered an overloaded circuit. A modern kitchen would have two separate dedicated circuits, one for the refrigerator and one for the microwave. Some older homes will have an inadequate number of circuits per room and this can lead to gang receptacles and potential circuit overloading.

There are meters available which will monitor and record any voltage drop. If a 10% voltage drop is present there may be an overloaded circuit or a poor connection. Often a motor such as a refrigerator compressor coming on will put a high demand on the circuit. Once the motor is running, the demand should drop. In either case, an electrician should be called in to repair the problem.

An inadequate 30 amp fuse panel.

Wire End View Fig. 3EL

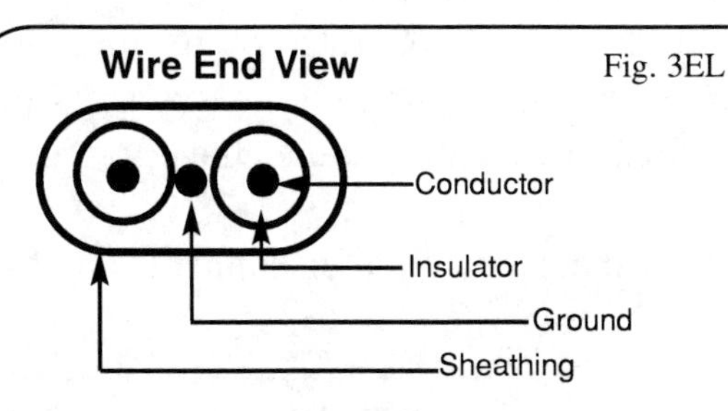

Interior Amps	Circuits Copper	Aluminum*
15	14	N/A
20	12	10
30	10	8
55	6	4

Low voltage, doorbell, thermostat takes from #18 to #22

*Single stranded aluminum wiring was used in the sixties and early seventies and is now considered a fire hazard. Proper connections must be made to prevent possible shorting and over-heating to the wiring. (See Aluminum Wiring) Larger multi-stranded wiring with an antioxidant cream on the conductors is acceptable.

⚠ *Problems to look for: ELECTRIC*

- Deteriorated wiring or sheathing
- Overhead power lines too close to ground, roof or deck
- Under sized service such as 30 amp or 120 volts, which is totally inadequate for a modern house.
- Two wire power lines (service drop)
- Mismatched service wire and panel box (main breaker or panel rating is different from service wire size rating.)

PANELS

The service wire runs to the meter via the masthead or an underground feed. From the meter, the service is then run to the main panel. The main panel is where the service wire is connected to the bus bar. These connections are called the lugs. If the wire is aluminum the connections should be coated with an antioxidation paste. The panel box contains overload protection devices such as fuses, breakers and/or disconnects.

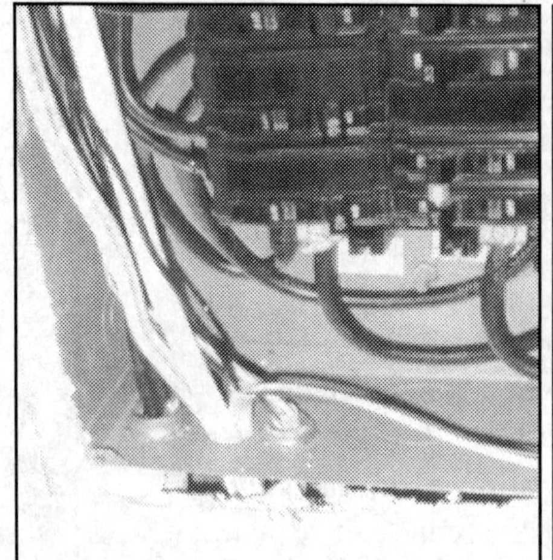

Sloppy circuit wiring missing connector at panel.

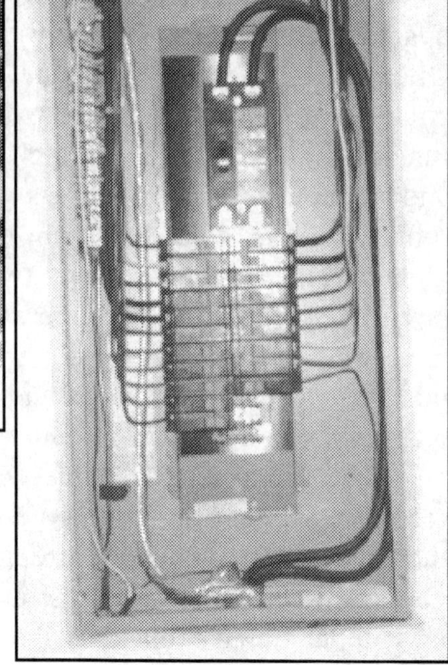

A professionally installed electric panel.

Modern electrical service usually has a main disconnect at the panel or by the meter. There are some panels with no disconnect capability, however, according to the NEC (National Electric Code) if eight movements of the hand can shut off all power to the house this is considered adequate. If not, a proper disconnect should be installed.

There are several manufacturers of panels, all of which have their own unique breakers. Older style pushomatic breakers may not be as readily available as modern breakers. Federal Pacific (FPE) STAB-LOK breakers have a history of not tripping properly, which could pose a potential fire hazard. Breakers should be routinely operated by turning them off and on. It is recommended that you wait a few minutes before turning on any breakers that service any compressor units such as a refrigerator. This will allow the pressure on the compressor to be released.

Fuse boxes will generally have a main fuse block that will act as a main disconnect. The fuse block has two cylindrical shaped fuses. Occasionally, someone will install a penny or aluminum foil in a fuse that continually blows. (Look for one fuse protruding past other fuses). This should be immediately removed since this is a fire hazard. If this has been an on-going problem the wire may have been subjected to overheating and caused cracking to the sheathing, allowing the metal to deteriorate. Copper wiring that has been overheated may have a reddish brown center and a tinned surface. At first glance the wire could pass as aluminum, until the surface is scraped away.

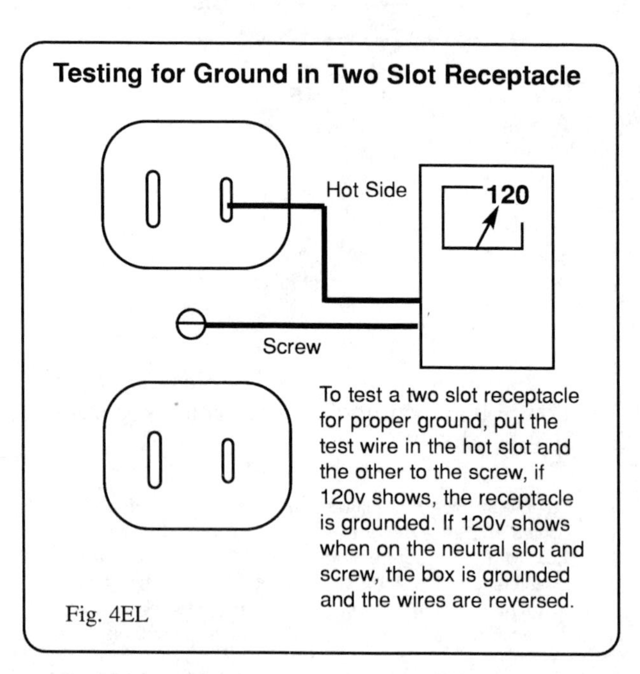

Fig. 4EL

When one or more sub-panels are run off the main panel, they should have a separate disconnect at the main panel or at the sub-panel. The wire size and the breaker for the

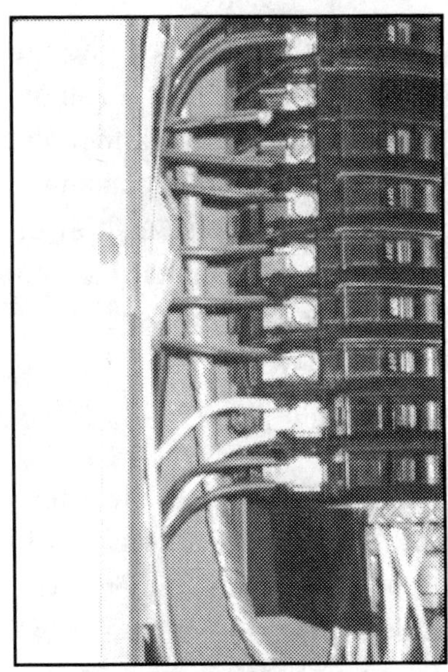

Double tapped circuit wires to breaker may overload breaker.

sub-panel should not exceed that of the main panel. If the main service is too small the sub-panel could demand too much of main panel and trip the main --- or heat up the wiring --- which could be a potential fire hazard. In this case, an electrician should be consulted, and upgrading the system may be necessary.

Panels are usually located in the garage, basement, and laundry room. In smaller and older homes they are often found in any interior room. If you cannot find the panel go outside and see where the meter and service wire are. From here you can determine an entry point. Usually, the panel is on the opposite side of the wall. Occasionally, you might have to trace the exposed circuits and look for the large quantities of runs in an attic or basement. They might lead you in the direction of the panel. Once you locate the panel it could be sealed in a wall or partially blocked, inhibiting access.

Look for a main disconnect for the panel. Most areas now require new homes to have one. Sometimes the disconnect will be located outside by the meter.

When inspecting the panel *only a qualified individual should attempt to remove the panel cover*, or electrocution and death could result. When inspecting the panel there should be a functional door on the front to protect the fuses. Also note if the cover is missing, this should be installed immediately to prevent possible electrocution. Check the face of the panel and see if any breaker knock outs are missing; these should be filled with blanks.

With the box door open, look for any excessive dirt, lint or other debris. The panel should be clean. See if there are any signs of rust or corrosion. Sometimes, rain will enter the house and trace the service wire, running into the panel. This can be prevented by sealing the wire at the point of entry.

When inspecting the wire, look for amateur or sloppy work. Look for double tapped breakers; this could lead to an overloaded circuit. Occasionally, a low voltage wire to a doorbell or furnace may be a double tapped wire. This is of no concern, however, any double tapped wire should be verified or separated.

Over-fusing is when a wire is too small for the fuse or breaker to which it is connected. This could allow overheating of the wire. For example, a 14-gauge wire connected to a 20amp breaker is considered over-fusing. The 14 gauge wire is only rated for a 15-amp breaker. If this has been an ongoing problem, the wiring may have been subjected to overheating causing cracking to the sheathing and allowed the metal to deteriorate.

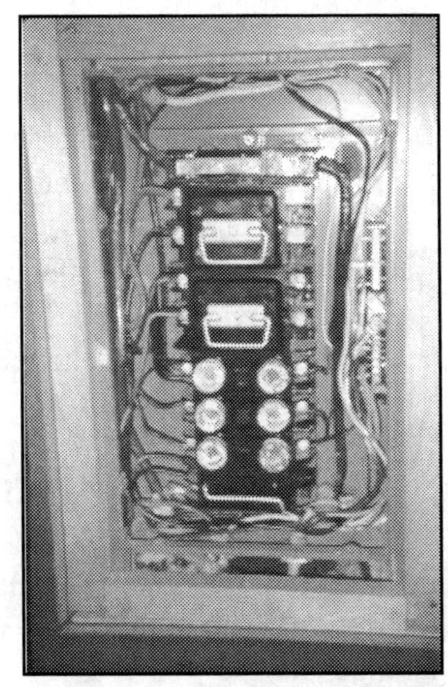

A fuse panel may be adequate as long as fuses rated for circuits are kept at original ratings.

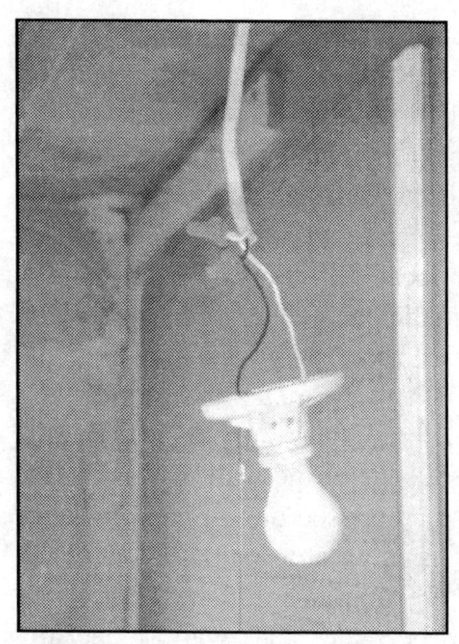

Light fixture should be properly secured.

Inspect the wire to box connections. They should have a positive connection to the box such as a romax connector. If the wire passes through the knockout hole, the sharp edge could cut the sheathing, causing a short; or the wire could be pulled and cause a short or break the connection. Note: Any panel or hard wired appliance, such as a garbage disposal, dishwasher light fixture, etc., must have a positive wire connection to the wiring box with a romax connector or other approved connection device.

It is not advisable to have a bed next to an electrical panel. Electromagnetic Field (EMF) hazards are still not defined, but electrical panes do emit higher EMF readings, so it is not advisable to spend long hours next to the panel. (See "Environmental" section on EMF).

 ***Problems to look for: ELECTRIC PANELS***

- Missing panel cover.
- Non-accessible panel, blocked, covered or concealed.
- Missing knock out plugs at panel cover.
- Missing romax or BX connectors.
- Double tapped wiring.
- Over fusing.
- Rust or corrosion.
- Nonfunctional breaker.
- Deteriorated wire sheathing.
- Disconnected or no ground.
- Sloppy wiring (amateur work).
- No disconnect.
- Excessively wet or damp area (such as a wet basement).
- Wire tapped off of the main lugs with no disconnect.

BRANCH WIRING

Much of the interior wiring is not visible for inspection. The attic, crawlspace, basement, and panels are some of the accessible areas where you may observe conditions. Any visible open splicing, sloppy wiring, or amateur workmanship may call for an electrician.

Electric wire connections should always terminate in fixture or junction box.

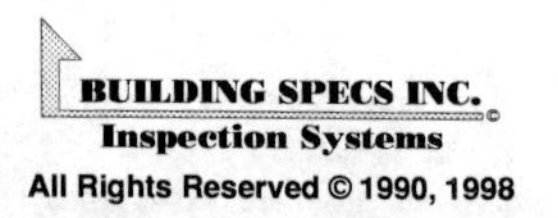

Burned or warm receptacles should be evaluated further as well.

There are several types of branch wiring for which you can look:

• *Romax* is a plastic sheathed wire commonly used in residential and light commercial construction. Non-metallic (NM) is listed for interior use. UF is rated for exterior and underground use, and will not breakdown in sunlight.

• *BX or metal armored wire* is commonly used in inner city and commercial buildings. The metal covering protects the wire from rodents chewing on the wire. Some older BX do not have a separate ground wire and the metal sheathing acts as a ground, this is not considered a proper ground. Some older BX have a rubber insulator that is prone to deterioration and shorting out against the sharp metal sheathing. In these cases, the insulators should be wrapped with electrical tape for extra insulation.

• *Conduit* is a metal or plastic tube that protects wiring.

• *Rag or cloth wiring* refers to an older wiring with only two wires and no ground. The insulators are an older rubberized wire (called loom) and it is prone to deterioration. When the sheathing (cloth) deteriorates (usually due to overheating of the circuits) the circuit is susceptible to shorting out at the switch boxes as well as the panel. This can also be a potential fire hazard, and should be updated by an electrician. Old BX can also be susceptible to these same conditions and should be further evaluated.

• *Knob and tube* is an antiquated wire. It is easily identified as it has push button switches and porcelain (knob) hangers for the wire. Though some consider it one of the safest types of wires, it is usually inadequate for modern house loads. Quite often rag wiring has been tape spliced directly to the ends of the knob and tube with no junction box. This type of wiring tends to emit a slightly higher EMF readings as well, and it is recommended to remove and update this type of wiring. Knob and tube has no sheathing over the insulators.

• *Aluminum wiring*: See special section on Aluminum Wiring at the end of this chapter.

OTHER ELECTRICAL PROBLEMS

- Open spliced wiring, missing junction box.
- Open junction box.
- Three pronged receptacles with improper ground
- No ground
- Reverse polarity
- Insulation in an attic over a non-rated for insulation contact recessed light (fire hazard).
- Lack of GFCI receptacle
- Non-functioning GFCI receptacle

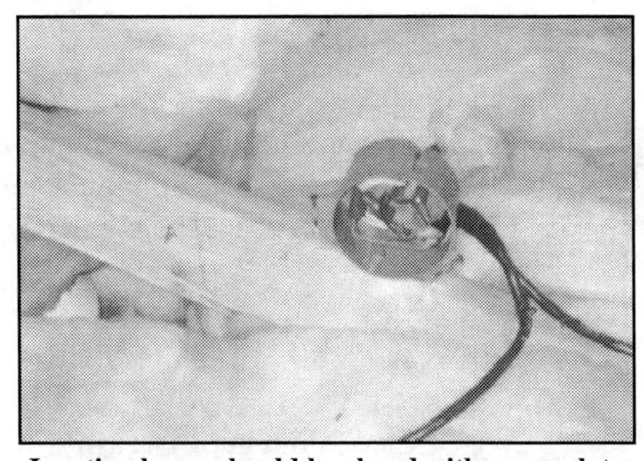

Junction boxes should be closed with cover plate.

A GFCI (ground fault circuit interrupter) is a safety device to protect from electrocution near a water source such as sink or outside receptacle. Basically a GFCI monitors current flow between

the hot and neutral. If there is a drop in the neutral the circuit is tripped. These should be routinely tested and can go bad. Once they fail they should be replaced.

 ALUMINUM CIRCUIT WIRING 

Solutions

Aluminum wiring was used in the 1970's for the wiring of receptacles and switches, but since then, single stranded branch aluminum wiring has been implicated in house fires. Aluminum does not conduct electricity as efficiently as copper and creates more resistance and heat. The wire also expands and contracts more than copper. The problem is compounded by screw heads too small at the devices. The expanding and contracting works the connections loose at the receptacles and switches. A short can occur as oxidation builds up between the aluminum wires, which poses a potential fire hazard. Updated devices with larger screws were created, however, these are not considered the best or only solution. Some people feel the aluminum should be removed and replaced with copper. Note that standard wire nuts are not approved for pig tailing and, according to the Consumer Product Safety Commission, may be an even greater fire hazard.

Approved Methods According to the Consumer Product Safety Commission:

- Copper pig-tails *crimped* onto the aluminum at all receptacles and switches. A special tool is used for this application.
- UL approved wire nut connectors, filled with anti-oxidant cream, designed specifically for aluminum to copper connections.
- Secure connections as is on approved Al/Cu devices.
- Antioxidant on all aluminum wiring connections
- Approved aluminum rated devices

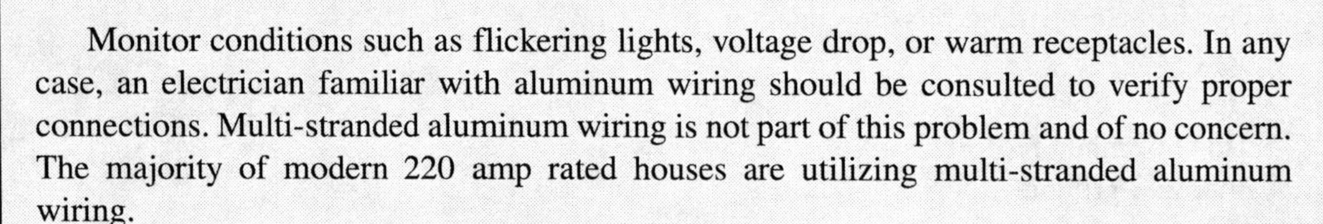

Monitor conditions such as flickering lights, voltage drop, or warm receptacles. In any case, an electrician familiar with aluminum wiring should be consulted to verify proper connections. Multi-stranded aluminum wiring is not part of this problem and of no concern. The majority of modern 220 amp rated houses are utilizing multi-stranded aluminum wiring.

COPPER CLAD ALUMINUM

Copper Clad Aluminum is easily mistaken for copper, and does not have the same problem with oxidation build-up as regular aluminum. It is typically a #12 wire and still needs proper devices with the larger screw heads or approved pig-tailing methods. To identify Copper Clad Aluminum look for a silver color at the ends of the wires where they are connected to the grounding bar. Another location to identify Copper Clad Aluminum is in the attic or crawlspace. In these areas look for the identification on the wire sheathing. Such as CU Clad AL.

Appliances

DISHWASHER

 *Problem:*

- Look for any water under the unit (broken hose, kinked or clogged drain hose or bad door seal).
- Excessive noise (foreign object stuck in pump, bad pump or motor), keep drain clear.
- Missing or overflowing air gaps. Commonly air gaps overflow due to grease solidifying in the discharge hose at the connection of the trap or disposal and needs to be cleaned out at this point.

GARBAGE DISPOSAL

 *Problem:*

- Excessive noise (foreign object stuck inside, bad motor). May need service or nearing end of life.
- Poor electrical connection, missing Romax connector, loose ground wire. Repair.

REFRIGERATOR

 *Problem:*

- Noisy (bad compressor or motor, dirty blower). Clean or have serviced.
- Water inside on bottom (clogged drain below bottom drawer, low refrigerant charge & not cooling properly, door was left ajar, blower dirty). Have serviced, keep air intake cleaned.
- Damaged door seal. Replace.

WATER HEATER

 *Problem:*

Electric

- No pressure relief valve (Hazardous, tank could explode, recommend installing one immediately).
- Pressure relief improperly directed (could expel at eye level or cause severe scalding,

recommend directing relief opening down near floor).
• Hot and cold lines reversed (could affect efficiency, have properly installed).
• Corrosion at top of unit (leak from above bleed nut, valve or shut of, can cause jacket to rust through).
• Open bottom access panel and visible corrosion, water, damp insulation (leak at element seal, leak from above connections running down, tank failing and rusting through). Have serviced and replace if necessary.
• Water on floor (leak at element seal, leak from above connections running down, tank failing and rusting through). Have serviced and replace if necessary.
• Water above 125° F (turn down at elements). This will prevent scalding and lower fuel cost.
• Water not heating (breaker turned off, element not working, bad relay, improperly wired) Have serviced.
• Unit setting on mud, dirt or wet area (can lead to rust and tank failure) elevate onto dry level platform.
• Finished floor level (install drain pan in case of tank failure).

Gas
• All that apply under electric unit.
• Poor flame shape or color at burner (needs to be serviced and cleaned yearly).
• Excessive rust piles on and around burner (needs to be serviced and cleaned yearly).
• Water streaks inside of the jacket (possible tank failure). Monitor for problems.
• Improper, missing or damaged flue. Flue should cover exhaust port at top of unit, slope up, and connect securely to any main flue. All connections should be male to female so as to not allow any combustion gases to escape into the living space.

WASHING MACHINE

 ***Problem:***

• Noisy (bad belt, motor failing). Have serviced.
• Low water pressure (well water is usually the cause, mineral deposits collect in the supply hose screens). Remove the hoses from the back, and clean any residue from the screens, and then replace them.
• Water on the floor (overflow, failing tank). May need to be replaced.
• Finished floor level and no overflow pan (install in case of an overflow, have a drain connected to carry water to a waste line).

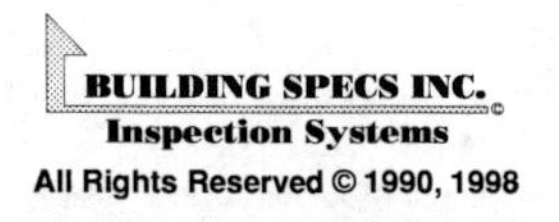

ELECTRIC CHECKLIST

1] **SERVICE ENTRY** ☐ **4/0** *(200 amp rated Al)* ☐ **2/0** *(150 amp rated Al)* ☐ **#2** *(100 amp rated Al)* ☐ **#4** *(60 amp rated Al)* ☐ **#6** *(60 amp rated Cu)*

☐ **Problem**[1]; Inadequate Service / Non Grounded System / Hazardous Condition, Location / Deteriorated Sheathing / ***Unknown Amp Rating***

2] **PANEL(S) MAIN #1 (*AMPS*______) MAIN #2 (*AMPS*______)** ☐ **Sub Panels (*AMPS*______)** ☐ ***Unknown Amp Rating***

☐ **Breaker** ☐ **Fuses** ☐ Ok ☐ **Problem**[1]; Double Tapping, Lugs, Circuits / [*Missing*, Cover, Knock-out Plugs, Wire Connector(s)] Non-Accessible / Burned Wires / Over Fusing, Inadequate Wiring Size / No Antioxidant on Aluminum Wires / No Visible Disconnect / FPE Panel[3] Could Not Remove Cover / Improper Grounds / Mismatched Service Wire and Main Disconnect, Breaker / *Amateur, Sloppy Workmanship*[1]

3] **BRANCH CIRCUITS** ☐ Ok ☐ Copper ☐ **Aluminum**[2] *(Requires UL approved connections be verified or installed by licensed electrician)*

a] **Style** ☐ Romex (NM) ☐ Rag / Cloth[1] ☐ Metal Armored (BX) ☐ Knob & Tube[1] *(Antiquated, non grounded)* ☐ Extension Cord[1]

☐ **Problem**[1]; Open Ground(s) / False Ground(s) / Reversed Polarity / Dead Circuit(s) / Burn Marks / Inadequate / Damaged Receptacles Aluminum Wiring[1] *(Potential Fire Hazard)* / Open Junction Box / Deteriorated Sheathing / Open Spliced Wires / *Amateur, Sloppy Workmanship*[1]

4] **GROUNDING** ☐ Ok ☐ **Footing** ☐ **Plumbing** ☐ **Rod** ☐ **Gas Line** *(Potentially Hazardous)*[1]

☐ **Problem**[1]; Spliced / Gas Line / False Ground / Plastic Interruption *(Non Conductive)* / No Ground / Severed or Disconnected /***Not Visible***

•**ADDITIONAL COMMENTS** *Inspection is for visible, exposed wiring only, and is limited to random sampling. New occupants may put different load demands on the electrical system which can in no way be anticipated. [1]Any repairs should be conducted by a licensed electrician. Rag, cloth wiring, Knob & Tube, or other should be replaced if sheathing is deteriorating, and are non grounded systems. Any evidence of burned or deteriorated wiring requires further evaluation by a licensed electrician. [2]Aluminum wiring has a history of being a potential fire hazard and approved connections should be made by a licensed electrician. [3]FPE Federal Pacific Panels (Stab-lok) have a history of failure of the breaker to trip properly.*

____

____

____

____

____

____

____

____

____

____

____

____

____

____

____

____

____

____

____

____

____

____

____

____

____

MAIN POWER ☐ **On** ☐ **Off** *(Could Not Evaluate)* ☐ ***Further Evaluation and Repairs Required by Licensed Electrician***[1]

Section Seven:

Interior

Room Surveys

When conducting a room survey, many of the inspection methods performed will be repeated in each room. You will want to look at all components including walls, ceiling, floor, doors, windows, receptacles, and check for proper function of various switches and items.

DOORWAYS

Water infiltration at exterior casing and toe kit may result in interior damage as well.

Inspect all door entries for the condition of the door and jamb. Verify the presence and condition of the weather stripping and seal. Look at the threshold and the sill, and feel the bottom of the door for the condition of the door sweep for its ability to keep water and air out.

As you close the door, look for any daylight that shows through and around the jamb. This could indicate a poor seal, which can affect your heating and cooling efficiency. The gap around the door should be fairly even, about a nickel's width at the top and along the striker plate side, and a dime's width at the hinge side.

Kneel down and tap the flooring along the threshold. Look for signs of rot or other damage to the subfloor or flooring. There might be a space at the bottom of the casing, if so, insert a probe into the subfloor. If the probe pushes in easily there may be damaged wood. Also look at the bottom of the casing for signs of water wicking due to leaking around the door frame. Non-destructive moisture meters can come in handy, as they can detect signs of moisture below the finished surface.

Inspect subfloor with probe for signs of water or insect damage.

Leaks which occur in these locations may be due to an

outside surface that is above, level, or slightly below the threshold. This, combined with an inadequate sealer or the lack of flashing below the threshold, can allow water to be absorbed by the edge of the subfloor. If there is a toe kick outside below the threshold, and it protrudes past the edge of the threshold, the same condition may occur. When ceramic tile or slate is present, moisture will sometimes cause tiles to become loose. These areas may be further investigated from below when access to a crawl space or unfinished basement is possible.

If water infiltration is apparent, further invasive investigation may be warranted. If the subfloor is showing signs of rot or termite damage, there may be a chance of damage to the floor joist, band board, wall studs, jacks, and bottom plate.

WALLS

At this point you will need a flashlight; the brighter the better. Turn off any lights in the room and hold the flashlight against the wall at one end of the room. Shine the beam down the wall towards the other end. The shadows that occur, will show you high spots, nail pops, sagging, cracks and even water stains. Hold the flashlight against the wall near the floor, and shine the beam up towards the ceiling and look for the same features. This procedure should be repeated on each wall, the ceiling and even some flooring (sheet goods). Glossier finishes will show surface irregularities even better. Nearly all walls and ceilings will show normal irregularities and imperfections.

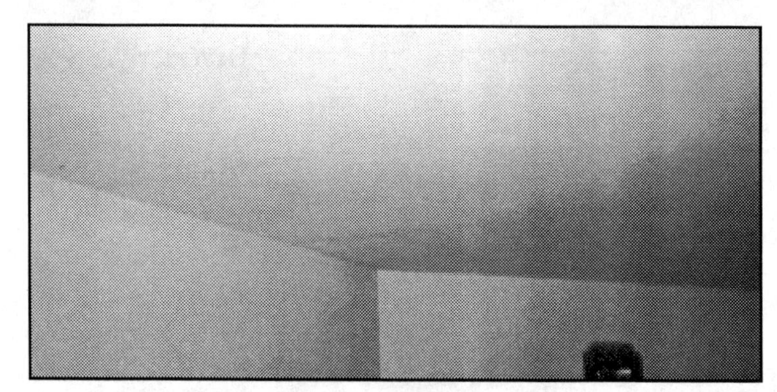

Look for cracking, irregularities, and water stains by utilizing a flashlight.

Drywall

Drywall, originally known as plasterboard, was designed as a backing for plaster, replacing lathing strips and wire mesh. A thin coat of plaster was applied directly over the plasterboard.

Drywall consists mainly of a gypsum core sandwiched between paper. It comes in various styles including a water resistant and a fire resistant form. Drywall comes in 1/4", 1/2" and 5/8" thickness. A paper or fiberglass tape embedded in joint compound creates a bond at the connections and inside corners that will not crack through the joint compound. Outside corners have a metal corner embedded in joint compound. The drywall may be screwed or nailed in combination with an adhesive. Application techniques may vary around the country.

Gradual, smooth, in and out "waves" may be associated with minor crowning and normal undulation of the studs and rafters. This can even change with heating and cooling. All buildings have a normal amount of expansion and contraction. A by-product of expansion and contraction may be nail pops; however, these may be attributed to nails that missed a stud or rafter. Drywall tape may start to lift due to a lack of joint compound between the tape and drywall. This may be evident by a dark line along the edge of the drywall tape. These are both usually just a cosmetic nuisance and can be easily repaired.

Drywall tolerates a lot of expansion and contraction; however, if cracks are apparent, the cause

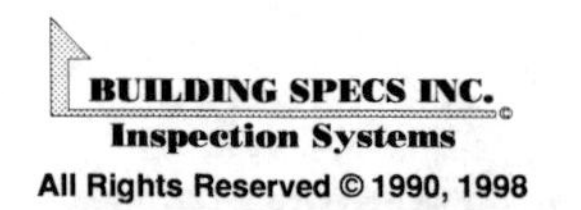

of their origin needs to be determined. Some cracks may be due to joint compound being used without drywall tape. Other cracks may be due to settlement or shifting of the wall. The direction of the crack may help you to determine the cause (refer to foundation settling).

Plaster

Plaster is a gypsum by-product spread as a finish coat over drywall or a browncoat backing. The browncoat is applied over wood or metal lathe to give the backing support. Some older homes may have animal hair in the browncoat to act as an aggregate and help hold the product together.

As the plaster finish coat ages, it may become loose, and separate from the backing. This can be determined visually or by tapping with your fingers, being careful not to damage any loose plaster. If the finish coat is in good condition you need to check the backing. Browncoat can deteriorate over a period of time and crumble and become loose from the lathing. Check for this condition by pressing firmly on the surface and note any signs of sponginess. Where this condition has become significant, it is common to see drywall installed over plaster, screwed through and into the framing. This can add extra weight to the building and cause the trim to lose some of its finished depth surface. When a plaster ceiling becomes loose from the backing, the possibility of the ceiling falling becomes an issue. Immediate professional attention and repair is required, as the weight of the ceiling falling on someone could certainly be fatal.

Wood

There are several styles and species of wood paneling and finishes. Sheets of wood or plastic veneer are the most common. Thinner sheets should be applied over a backing such as drywall; the sheets should be glued and nailed. When a backing is not present, the paneling is susceptible to severe undulation in between the studs.

Wainscot and other vertical woods should be applied over a backing such as solid plywood or several rows of horizontal blocks to allow for a proper nailing backing.

There are many other wall finishes available. The same general inspection methods should apply to all of them.

CEILINGS

Signs of water stains should

When inspecting the ceiling, look for any sagging or loose sections. You may want to set up a ladder, push up on the ceiling, and look for drywall or plaster separation. If water stains are found, trace them back to the source if possible. If there is an attic access over the stain, start there. Have someone tap the ceiling as you locate the stain. You may have to pull some insulation back. From that point check the roof sheathing and framing above the area for any stains or damp spots. If there is a valley intersection or roof penetration such as a chimney, plumbing vent, or skylight, look for signs of water infiltration.

Once you think you have located the source get on the roof (if possible) and try to seal the source. If you do not suspect the roof of leaking, look at the plumbing including fire-sprinkler (where applicable). Another source for water problems can be a plumbing vent with a bad connection, or a sagging dryer vent. Both of these can trap large amounts of condensation.

As you enter a room look at the condition of the door casing and jamb. Check to see if the striker plate is installed at the door jamb and latches properly. Close the door, and look for any binding, which could indicate minor settling or poor installation. Check the door's surface for any damage. As you enter the room, check the lock (if applicable) and examine the condition of the trim inside. Note whether there is a door stopper. Without one, the knob may damage the adjacent wall.

If the top of a door frame or any hall entry is severely out of level, or a door will not close, further investigation may be necessary. This could indicate settling elsewhere in the house and the problem area may need to be stabilized.

Check all windows and locks for proper operating condition. They should function smoothly and stay in an open or closed position as needed. Look around the frame for any visible signs of moisture. Inspect the bottom of the sill, especially when constructed from wood. Condensation collecting on the window pane can run onto the sill and cause water damage. Check the glass for any cracks and note the presence and condition of the screens. In basements and bedrooms with no doors, proper egress for fire escapes should be considered.

Check all closets for door operation and damage. See if the hanging pole is properly supported.

ELECTRICAL

There is usually a group of switches at the entryway or foyer. Turn them on one at a time, and by process of elimination, you can determine which operates what.

As you walk around, take note of any recessed lights. Once you get in the attic, verify whether or not they are rated for insulation contact. Test all of the accessible receptacles with a receptacles tester. A basic tester will show if there is an open ground or reverse polarity; some will test for a ground fault interrupter. There are more sophisticated testers which record any voltage drop, while others will induce a load to test for capacity. If there is a switched receptacle with no light, turn the switch off and using a receptacle tester, go to each receptacle until you find one that is off. Then turn the switch on to check for proper operation.

If there is a ceiling fan, turn it on, and let it run in all speeds and modes. Look for severe wobbles or a poor connection at the ceiling. Listen for any noise which could indicate bad bearings. A minor wobble could be dust on the paddles throwing the fan out of balance.

HEATING

Verify the presence and condition of a heating source. When forced air is used, and there is no separate air return, be sure there is a gap between the door and the floor. When the door is tight to the flooring, it causes incoming air to slow down if there is nowhere for the existing air in the room to escape. Once the room becomes "pressurized," heating and cooling becomes less efficient.

FLOORING

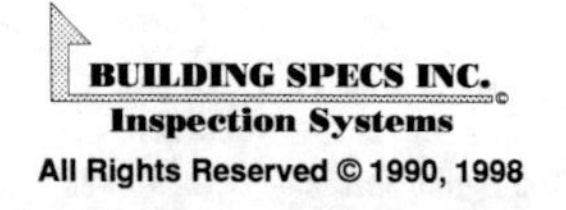

As you inspect the condition of the flooring, bounce lightly to feel how much deflection there is in the structure. If it seems excessive, additional support or further structural evaluation, may be required. The flooring is considered a cosmetic item, and is not covered in an average inspection unless the surface is severely damaged.

There are many types of flooring:

Carpet comes in many styles and grades, some are much more resistant to stains and wear than others. Inspect the carpet for heavy traffic wear areas and deteriorated weave. Some carpet may emit what is referred to as Volatile Organic Compound (VOC's). Volatile Organic Compounds may also be given off from new paint, varnish and other finishes, and typically dissapates over time.

Hardwood flooring comes in differing styles ranging from narrow, to assorted planking or strips, to parquet tiles. The hardwood may be a solid piece of wood or a thick veneer over a plywood substrate. Solid wood is available in various species including oak, fir, southern yellow pine, clear heart pine, and many others. There are a few standard finishes associated with hardwood floors.

One style is a pre-finished surface. Pre-finished floors can come in various thicknesses ranging from 3/8" to 3/4". The finish may be either a urethane or acrylic impregnated finish. The flooring may be nailed or glued down, preferably both. Edges may be beveled down or flush. Some of these floor will not allow re-sanding, so check with the manufacturer.

Solid hardwood should be installed over a rosin paper and nailed together tightly. After the flooring is installed the surface is sanded and sealed. When inspecting this type of floor, especially in an older house, look for exposed toe nails. If the floor has been sanded too much, the tongue and groove can be exposed, as well as the nails. This can lead to loose boards and the inability to resand. Also, look at the perimeter for "dishing" in the wood which can indicate the extent of sanding from the original surface level.

Look for signs of loose or squeaky boards. Large gaps may indicate the flooring was installed with a high moisture content, and once heat was introduced into the house, the boards became dry and shrank. Another sign of existing moisture (especially at door entries) is cupping in the wood. Moisture wicking into the wood may not stain or be visible, but it will swell the wood enough to force the edges up. Simply run your hand across the boards, or shine a bright flashlight across the boards to see this problem.

Sheet goods (linoleum, vinyl flooring) come in several styles and qualities, from a thin vinyl coated material to a thick color embossed one. The seams can be glue welded together to hide repairs or joining. Most manufacturers require that the flooring must be laid over a smooth clean surface. Any sharp or raised edges from the subfloor could wear through the material prematurely. The flooring is applied over a thin mastic coat applied with a 1/16" notched trowel. If the adhesive is applied too heavily, the floor may have soft pockets or bubble up.

Linoleum tiles also come in various qualities and thicknesses. Some tiles must be applied over a thin mastic coating to the subfloor, while others have an adhesive on the back (peel and press). The self adhesive tiles tend to not hold up as well, and may pop loose. Older tiles and mastic may contain asbestos. If asbestos is suspected, a small sample may be taken for testing from an inconspicuous area, such as a closet or under a staircase. Refer to the "Environmental" section for more information.

Ceramic Tile, Slate, and Marble have some similar requirements and problems. Most manufacturers require a very stable and solid subfloor, such as concrete, concrete board, or 1" to 1- 1/4" of solid wood. For example: 3/4" plywood sheathing and a 1/4" A.C. subfloor or more can create this subsurface. If tiles are laid on an unstable surface, the tiles may pop loose or even crack. Shine a bright flashlight across the tiles to look for severe floor deformities or cracks. Marble hides cracks in the veins' colors, so if you suspect a crack, run your fingernails across the crack in question and feel for a sharp edge. To check for loose tiles, tap the tiles lightly with your fingers and listen for a hollow or clicking sound. If the tiles are good the tapping will produce a solid sound. Also check the integrity of the grout by scraping it with your fingernail. A loose or sandy consistency could be a sign of failure, and may need to be pointed up.

Floor Structure

The floor structure itself may have sagging, high spots or other deformities. Some irregularities may be normal, but may also be a sign of a more severe problem.

Some problems to look for: INTERIOR ROOMS

• A wall which runs parallel with the joist, may push down on the sheathing, (plywood) and sag in-between the joist. This can be prevented by running one joist offset from below the wall, or have two joists spaced the width of of the wall (to allow for a plumbing/mechanical chase). Another method of prevention is to install solid blocking at a perpendicular angle to the joist, approximately 16" on center, thus creating a ladder effect.

• Look for any beams that have joists intersecting along one or both sides which have inadequate connections, such as missing joist hangers. Joist that have begun to sag over a period of time may create a high spot at the beam. If the condition is severe, the joist may need additional support.

• Excessive span, or a lack of bridging can lead to the joist sagging. If there appears to be a problem refer to local or national span load tables. Remember, the wood's grade and species will affect the span. The joist spacing will also affect the span allowance. As you walk around, listen for the rattle of object in any cabinets or, bounce lightly to feel for excessive bounce. One simple test to measure the amount of deflection is to use a telescoping (friction type) paint pole. Loosen the friction slightly, hold the pole plumb, letting one end rest on the floor, while the other end touches the bottom of the floor joist. Then have someone who weighs about 175 lbs bounce on the floor above you. As the pole slides together, mark where the poles bypass each other. Repeat this procedure a few times to determine an average. If the deflection is over an inch, further structural evaluation may be necessary.

• Rooms may have been added on at different times and thus created a floor which has poor transitions. While this may not be considered a structural problem, it could be a nuisance.

• Ballooned framed walls are notorious for failing at the wall connections. A problem may be hard to detect until it is too late and the floor breaks loose. In some cases a ledger or a sistered wall may be necessary to support the floor.

• Floors built directly on top of the ground, or built too close to the ground, may be susceptible to water and insect damage. Inspection of these areas is usually limited, unless sections of the floor are removed.

• Severe weak or sagging areas, doors binding, door frames sagging, etc., could be a sign of

severe structural failure. These problems could be due to poor construction methods, inferior materials, unresolved water problems, excessive weight load (water bed, hot tub, bookcase), rotted wood, failing wood, foundation failing, or a multi-level addition added on top of the structure without proper structural reinforcement, etc. Further structural evaluation may be needed, including invasive inspection, such as wallboard or siding removal.

Joist and beam material can vary, depending on region, span, load, availability of material, builder's discretion, codes, and cost.

These are some common types of materials used:

- Dimensional lumber
- Plywood (I Beams)
- Micro Lams (Plywood veneers)
- Paralams (Oriented strand beam)
- Steel Truss
- Floor Trusses (2x4 triangulation web)
- OSB (I Beams)
- Glue Lams (2x4 glued up flat)
- Steel (I Beam)

A concrete subfloor may not be visible for inspection if there is a floor covering. You may be able to notice rolls and other imperfections; however, an inspection of the concrete itself is not possible.

When the slab is at, or near grade level, (including in a converted garage) check to see if the framed wall rests on the slab without an additional elevated masonry knee wall. If the framed wall is at, or near grade, a close inspection of the interior floor and outside wall base is in order. Look for water stains on the wall, carpet, or other interior floor finishings adjacent to the wall. Feel under the lip of the siding for water or insect damage to the sheathing. This area is subject to water or insect damage.

WINDOWS

Windows are available in several styles, makes and materials. Some windows are installed with new construction, while others are used as retrofit such as replacement windows. One type of replacement window is especially designed for older wooden double hung types. These windows allow a retrofit without disturbing the adjacent trims or casings.

Some window styles include:

- Double Hung: slides up and down, sashes are held in an open position by either friction, positive stops or counter weight.
- Gliding: slides horizontally.
- Casement: hinged on one side and cranks open.
- Awning: similar to a casement, hinged at top.
- Fixed: non-opening.
- Custom: built to different shapes and sizes.

Storm windows need drain holes or water will pond and rot sill.

The materials windows are manufactured

from can also vary:

- Wood
- Vinyl Clad
- Aluminum Clad
- Extruded Aluminum
- Vinyl (or other plastic)

Check window for square by lowering sash and observing reveal left between sash and window.

As you go through each room, test the windows by opening and closing them. If the windows are stuck, do not force them, or you may end up with a broken sash on your hands. Sticking windows may be attributed to high humidity, being painted shut, poor installation, pin locks through both sashes, settling, or poor maintenance.

Check for any cracked or broken panes, and note if there are any missing screens. If the windows are double pane look for condensation or a fogged appearance to the glazing, which could indicate a broken seal. Some windows have the make and year in the corner of the pane. If the windows are single pane, look for any signs of condensation damage to the sill or sash. If there are storm windows, verify that the drain holes are clear at the bottom. If not, water will not drain out when the screens are in.

Examine window locks for proper installation and function. If the windows are the only form of egress, note the size of the window opening. In the event of a fire, this may be the only escape from a room.

Look for any water stains below or around the window, which might indicate a leak. Check the window's squareness for signs of settling. Observe whether or not the window will stay open, or if it closes by itself.

MILDEW

As you walk through the house, note any musty odors or visual signs of mildew. This could indicate a moisture or dampness problem. The source should be located and inhibited. A dehumidifier will help to remove moisture from the air. A dampness problem may be associated with a failing heat system, which is allowing combustion fumes to enter thru the ductwork. In the event this is the case, immediate service to the heating system is in order.

INTERIOR CHECKLIST

1] **OVERALL CONDITIONS** ☐ **Further Evaluation and Repairs Required**[1] ☐ **Water Stains**[1] ☐ **Settlement Noted**[1] ☐ **Odor Noted**

2] **FLOOR** **Structure**[1] ☐ Ok ☐ **Problem**[1]; Bouncy / Sagging / Settlement / Hump / Severely, Visibly Out of Level / Rot / Squeaking

a] **Carpet** ☐ Ok ☐ **Problem;** Heavy Wear / Stains / Burn Mark / Mildew at Perimeter / Rippled / Deteriorated / Damp

b] **Hardwood** ☐ Ok ☐ **Problem;** Heavy Wear / Stains / Cupped Planks ***(Possible Moisture)*** / Squeaking, Loose Boards / Cracking

c] **Sheet Vinyl** ☐ Ok ☐ **Problem;** Heavy Wear / Stains / Burn Mark / Torn / Sub-floor Wearing Through / Loose / Curling

d] **Vinyl / Linoleum Tiles** ☐ Ok ☐ **Problem;** Heavy Wear / Stains / Burn Mark / Torn / Loose Tiles / Cupping Tiles / Cracked Tiles

e] **Ceramic, Marble** ☐ Ok ☐ **Problem;** Heavy Wear / Cracked Tiles / Loose Tiles / Amateur Installation / Grout Failing, Cracking

3] **WALLS** ☐ **Drywall, Plaster** ☐ **Wood** ☐ **Other**____________________ ☐ *Sloppy, Amateur Finish or Installation*

☐ **Problem**[1]; Settlement Cracks / Water Stains / Nail Pops / Holes / Separating From Structure / Bulge / Undulation Noted at Wall / Mildew

4] **CEILING** ☐ **Drywall, Plaster** ☐ **Wood** ☐ **Other**____________________ ☐ *Sloppy, Amateur Finish or Installation*

☐ **Problem**[1]; Settlement Cracks / Water Stains / Nail Pops / Sagging / Separating From Structure / Undulation Noted at Ceiling / Mildew

5] **WINDOWS** ☐ Ok ☐ **Problem**[1]; Inoperable / Sticking / Cracked Glass / Water Damage / Frame Out of Level *(Possible Settlement)*[1]

Window(s) Blocked by; *Furniture, Plants, Shades, Etc.* / Key Locked Closed / **[Lock(s);** Missing, Damaged Inoperable] / Won't Stay Open

6] **DOORS** ☐ **Hinged** ☐ **Bi-fold** ☐ **Pocket** ☐ **Sliding** ☐ **French** ☐ **Mirrored** ☐ **Other**________________

☐ **Problem**[2]; Inoperable / Binding / [Damaged Door, Frame, Jamb, Casing] / Not Latching / No Lockset / Off Track / Missing Striker Plate

Dragging on Carpet / Missing Wall Bumpers / *Settlement Observed at Top of Jamb*[1] / **[Door(s);** Blocked, Locked, Not Accessible]

7] **RECEPTACLES** ☐ Ok ☐ **Problem;** Non Grounded / False Ground / Reverse Polarity / Non Functional / Loose / Painted Shut / Damaged

Filled With Safety Plugs / Missing Plates / Burn Marks / Hazardous, Improper Location / *Blocked, Limited or No Access Due to Furniture, etc.*

8] **LIGHTS** ☐ Ok ☐ **Dimmer** ☐ **Switched Receptacle** ☐ **Ceiling** ☐ **Wall / Sconce** ☐ **Other**________________

☐ **Problem;** Missing Plate / Loose, Improper Installation / Lights Flickering, Possible Short / Damaged / Not Functional / None Located

9] **CEILING FAN(S)** ☐ Ok ☐ **Problem;** Wobbles / Noisy / Inoperable / Loose, Damaged Blade / Inadequate Connection / Hazardous Operation

10] **HEAT** ☐ Yes ☐ **Problem;** Trim Doors For Air Return / Drapes on Heater / Potential Fire Hazard / *Not Supplied in All Rooms*

11] **STAIR RAILING** ☐ Ok ☐ **Problem;** Loose / Too Low / Missing / Inadequate / Hazardous / Damaged / *Sloppy, Amateur Finish or Installation*

12] **TRIM** ☐ Ok ☐ **Wood** ☐ **Drywall, Plaster** ☐ **Vinyl Covered Wood** ☐ **Plastic** ☐ **Metal**

☐ **Problem;** Loose / Missing Sections / Inadequately Installed / Amateur, Sloppy Workmanship / Damaged / *Water Stains at Baseboard*

•ADDITIONAL COMMENTS *Interior room surveys do not include cosmetic defects, i.e., paint, trim or other finishes. [1]Any settlement or signs of moisture should be monitored for further activity. Any repairs or evaluations should be conducted by a licensed or qualified individual.*

Foyer: ____

____

Living Room: ____

____

Dining Room: ____

____

Master Bedroom: ____

____

Bedroom: ____

____

Bedroom: ____

____

Den, Office ____

____

DOORS & WINDOWS & ENERGY CONSERVATION CHECKLIST

Doors

☐Caulk Needed Outside, Around Door Frame ☐Inadequately Installed ☐Amateur, Sloppy Installation

1] ENTRY(S) ☐Single ☐French ☐Swingset ☐Sliding ☐Sidelights ☐Screen Door Missing

a] Material ☐Wood ☐Metal ☐Fiberglass ☐Vinyl Clad ☐Aluminum Clad ☐Vinyl

b] Condition ☐Good ☐Average ☐*Poor;* Rotted / Delaminating / Falling Apart / Cracked Glass / Bad Seal, *(Trapped Moisture)*

☐Problem; Inadequately Hung / Not Locking Properly / Lockset Damaged or Missing / Inoperable / Binding / Blocked / Loose At Hinges Glass Damaged / Jamb Damaged / Not Sealing Properly / Weather Strip Missing, Damaged / Not Exterior Rated Door / Screen Missing, Damaged

☐*Rot Noted At;* Jamb / Casing / Toe Kick / Subfloor / Door / Headtrim ☐Minor ☐Severe ☐*Repair/Replace As Needed*

•ADDITIONAL COMMENTS *Exterior caulks and sealants must be maintained to prevent water infiltration and damage.*

____

2] STORM DOORS ☐Yes ☐No ☐Recommended ☐Wood ☐Metal ☐Vinyl ☐Other______________

☐Problem; Sagging / Sticking / Damaged / Screen Torn / Cracked Glass / Poor Condition / Inoperable / Not Locking / Poorly Installed

☐*Missing or Damaged;* Closer / Safety Chain / Locks / Proper Seal / Miscellaneous Hardware ☐*Repair/Replace As Needed*

•ADDITIONAL COMMENTS

____

Windows

☐*Put Windows In Proper Operating Condition; Few, Several, Majority, All* ☐Caulk Needed Outside Around Frame

3] STYLE(S) ☐Double Hung ☐Casement ☐Awning ☐Gliding ☐Picture ☐Jalousie ☐Site Built

a] Frame Material ☐Wood ☐Vinyl Clad ☐Aluminum Clad ☐Vinyl ☐Aluminum ☐Tilt Pack

b] Glass ☐Single Pane ☐Double Pane ☐Triple Pane ☐Storm Window(s) ☐Plastic Wrap ☐Replacement

4] OVERALL CONDITION ☐Ok ☐Poor ☐Repairs Required ☐*Updating Recommended* ☐Amateur, Sloppy Installation

☐Problem; Cracked Glass / Broken Seal, *(Trapped Condensation)* / Inoperable / Won't Stay Open / Locks Missing, Damaged / Binding / Sash Cords Cut Frame Loose, Separating / Screens Missing, Damaged / Rotted Sill, Frame / Sash Separating / Glazing Putty Failing / Return Springs Damaged

•ADDITIONAL COMMENTS *Exterior caulks and sealants must be maintained to prevent water infiltration and damage.*

____

____

____

Energy Conservation

☐*Updating Recommended*

5] THERMAL PANE ☐Yes ☐No ☐Single ☐Double ☐Triple ☐Replacement ☐High Efficiency Glass

6] STORM WINDOWS ☐Yes ☐No ☐N/A ☐Recommended ☐Clogged Drain Holes ☐Plastic Over Windows

7] DOORS Insulated ☐Yes ☐No •STORM DOORS ☐Yes ☐No ☐Recommended *(Remove glass panel during summer)*

8] WEATHER STRIPPING ☐Ok ☐Problem ☐Doors ☐Windows ☐*Updating Recommended*

9] EXTERIOR CAULK ☐Ok ☐Problem; Missing / Cracking / Failing / Blown Out, Possible Water Infiltration

10] INSULATION ☐*Updating Recommended*

a] Floor ☐Yes ☐No ☐N/V ☐Inadequate ☐N/A Approximate R Value________ Depth________"

b] Walls ☐Yes ☐No ☐N/V ☐Inadequate ☐N/A Approximate R Value________ Depth________"

c] Ceiling ☐Yes ☐No ☐N/V ☐Inadequate ☐N/A Approximate R Value________ Depth________"

11] WATER HEATER Temperature*_______° ☐*Recommend Turning Down** Insulated ☐Yes ☐No ☐*Fire Hazard*

12] DUCT WORK Insulated ☐Yes ☐No ☐Recommended ☐Falling Apart ☐Restricted Air Flow

13] ATTIC VENTILATION ☐Ok ☐Problem ☐*Insufficient, Updating Recommended* *(Shingles may prematurely fail)* ☐Soffit Blocked

•ADDITIONAL COMMENTS **Lower the hot water heater temperature to help reduce the chance of scalding and burns to 125° F. This will also help you save money on energy.*

____

BUILDING SPECS INC.©
Inspection Systems

KITCHEN & APPLIANCES CHECKLIST

Kitchen

☐Multiple Kitchens[1] # of_____ *(Are incorporated into one report.)*

1] **OVERALL CONDITION** ☐**Adequate** ☐**Heavy Wear** ☐**Out Dated** ☐**Amateur, Sloppy Workmanship** ☐**Incomplete**

2] **CABINETS** ☐**Solid Wood** ☐**Wood Veneer** ☐**Formica** ☐**Metal** ☐**Plastic** ☐**Other**____________________

a] **Finish** ☐Clear ☐Stained ☐Painted ☐None ☐**Heavy Wear** ☐**Water Damage, Stains**

☐Ok ☐**Problem;** Poor Alignment / Loose / Damaged / Frames Falling Apart / Sloppy Installation / ☐**[Door(s), Drawer(s),** Loose, Binding]

3] **COUNTER(S)** ☐Laminate ☐Corian*(Or Similar)* ☐Wood ☐Ceramic ☐Copper ☐Metal ☐Marble, Granite

Condition ☐Ok ☐**Problem;** Loose / Delaminating / Cracked / Burned / Chipped / **Heavy Wear / Sloppy Installation / Water Damage**

4] **SINK(S)** **Material;** ☐Stainless ☐Steel, Enamel ☐Corian*(Composite)* ☐Ok ☐**Problem;** Loose / Heavy Wear / Damaged

a] **Faucet(s)** ☐Ok ☐**Problem;** Drips / Leaks / Loose / Damaged / Low, Sloppy Pressure / Poor Condition / Not Operational, Connected

b] **Spray** ☐Ok ☐**Problem;** Drips / Leaks / Loose / Damaged / Low, Sloppy Pressure / Poor Condition / Not Operational, Connected

c] **Drain(s)** ☐Ok ☐**Problem;** Clogged / Slow / Leaks / Amateur, Sloppy Installation / Corrosion / Loose / Failing / No, Limited Access

d] **Disposal** ☐Ok ☐**Problem;** Clogged / Not Operational / Noisy / Older Unit / Loose / Hazardous Wiring / Amateur, Sloppy Installation

5] **FLOOR** ☐Ok ☐**Sheet Goods** ☐**Vinyl, Linoleum Tile** ☐**Ceramic, Marble Tile** ☐**Hardwood** ☐**Carpet**

☐**Problem;** Torn / Curling / Cracked / Burned / Water Damage / Loose Tiles / Water On Floor / Amateur, Sloppy Installation / **Heavy Wear**

6] **RECEPTACLES** ☐Ok ☐**Problem;** None / Not Operational / Damaged / **[GFCI**[2] ☐Yes ☐No ☐Not Operational ☐*Recommend*]

•**ADDITIONAL COMMENTS** *[1]Multiple kitchens are incorporated into one. [2]A ground fault interrupter is recommended near all water sources to prevent electrical shock. Cabinetry not inspected for cosmetic defects, missing hardware, or proper operation. Some cabinets are poorly built and may not maintain their serviceability.*

____

____

____

____

____

____

Appliances

☐***SERVICE REQUIRED ON***____

1] **Exhaust Fan** ☐Yes ☐N/A ☐**Problem;** None / Recommended / Not Operational / Noisy / Missing Filter / Light Missing, Not Operational

2] **Dishwasher** ☐Yes ☐N/A ☐**Problem;** Not Operational / Water Under Unit, On Floor / Air Gap Overflows / Bad Seal / **Older Unit**

3] **Refrigerator** ☐Yes ☐N/A ☐**Problem;** Not Operational / Ice-Maker Empty / Bad Seal / Noisy / Damaged / Condensation / **Older Unit**

4] **Oven** ☐Ok ☐N/A ☐**Electric** ☐**Gas** ☐**Propane** ☐**Convection**

☐**Problem;** Not Heating / Not Igniting / Damaged Unit / Heavy Wear / Not Operating Properly / Element Burned Out / Knobs Missing / **Older Unit**

5] **Range** ☐Ok ☐N/A ☐**Electric** ☐**Gas** ☐**Propane** ☐**Downdraft** ☐**Grill**

☐**Problem;** Not Heating / Not Igniting / Damaged Unit / Heavy Wear / Not Operating Properly / Element Burned Out / Knobs Missing / **Older Unit**

6] **Microwave** ☐Yes ☐N/A ☐***Not Tested*** ☐**Problem;** Not Heating / Heavy Wear / Not Operating Properly / **Older Unit**

7] **Compactor** ☐Yes ☐N/A ☐***Not Tested*** ☐**Problem;** Not Cycling / Heavy Wear / Not Operating Properly / Leaking / **Older Unit**

9] **Washer** ☐Yes ☐N/A ☐***Not Tested*** ☐**Problem;** Not Operating Properly / Noisy / Needs Over Flow Pan / Rusted / **Older Unit**

10] **Dryer**[1] ☐Yes ☐N/A ☐***Not Tested*** ☐**Problem;** Not Operating Properly / Noisy / Not Heating / Improperly Vented / **Older Unit**

11] **Central Vac.** ☐Yes ☐N/A ☐***Not Tested*** ☐**Problem;** Not Operating Properly / Missing Components / Heavy Wear / **Older Unit**

•**ADDITIONAL COMMENTS** *Appliances are checked for operation, and no life expectancy or warranty is implied or given. An older unit may be nearing the end of it's useful life. Appliances should be monitored and maintained routinely. [1]Recommend a metal dryer vent, vinyl may be a potential fire hazard.*

____

____

____

____

Section Eight:

Environmental Concerns & Hazards

Asbestos

Asbestos is one of the best known contaminates, and commercial and residential property owners should be aware of both health and legal concerns regarding this notorious substance. Asbestos is actually a group of six different minerals (actinolite, amosite, anthophyllite, chrysotile, tremolite, and crocidolite) which occur naturally in the environment. Asbestos appears as long fibers, something like fiberglass, and due to their resistance to heat and chemicals, enjoyed an extended period of popularity as a building material. However, although asbestos' negative coverage would lead most people to believe that standing anywhere near an asbestos shingle is potentially harmful, asbestos is not actually a serious detriment to human health unless it has begun to crumble and is subsequently inhaled.

Asbestos wet rap on duct work may infiltrate air supply.

Low levels of asbestos can be detected in almost any air sample, but it is in enclosed areas where exposure to deteriorating insulation, for example, causes the most damage. Once asbestos shingles, insulation or other materials begin to breakdown, tiny fibers are released into the air. People begin breathing in concentrated amounts of these fibers, which then lodge in the lungs. Although not all fibers inhaled will stay permanently in the lungs, some may accumulate.

The U.S. Department of Health and Human Service has determined that asbestos is a known carcinogen. Exposure to increased levels of asbestos can cause cancer of the lung tissue or mesothelioma, a cancer of the thin membrane which surrounds the lungs and other internal organs. Unfortunately, these diseases are not only fatal, but they also develop slowly. Lack of immediate symptoms can keep people from realizing adverse affects until it is too late.
Other detrimental affects possibly caused by asbestos include the accumulation of scar-like tissue in the lungs, which hampers breathing, and possibly cancer in other parts of the body.

The U.S. Environmental Protection Agency (EPA) has banned most of the manufacture, processing, importation, and distribution of materials or products that contain asbestos. Legislation against the use of asbestos initiated in 1990 is currently in full force. The ban eliminates asbestos in insulation, brakes, floor and ceiling tiles, cement, paper, and nearly all other materials. The EPA has also passed many laws which monitor and control asbestos in public work places, buildings, and, in particular, schools.

How much a person is affected by exposure to asbestos depends on the length of exposure, age, health and other factors, but it is best to avoid all exposure if possible. Any known asbestos items in or on a home should be inspected for deterioration, and the air can be tested by a professional asbestos testing company. If levels in the building are dangerous, steps should be taken to remove or cover up the areas so that loose fibers cannot escape into the air.

Buried Oil Tanks

Statistics indicate as many as 25% of all Underground storage tanks (USTs) may now be leaking. These underground tanks may hold oil or gas but can most frequently be found at homes heated with oil. Over time, there is a tendency for these older tanks to corrode and leak their contents into the ground and possibly into the ground water. Because of the great expense involved in cleaning up the contamination caused by leaking tanks, it behooves those that have older underground tanks, to have them or the soil tested.

Because most underground oil tanks are steel, this steel can rust over time allowing the oil or gas to leak into the ground. Although it is hard to pinpoint exactly what age this may occur, the state of Maryland has stated that any tank over the age of fifteen may be at risk of leaking. It is recommended that any tank over twenty to twenty five should be checked to insure it's integrity.

If a tank begins leaking its contents into the soil the owner of the tank (house where tank is located) whether a new owner or long time resident may be responsible for the cleanup. It is important to remember that the only time that cleanup is required is when the oil or gas actually contaminates the ground water or public waterways. Obviously, those with high water tables or in close proximity to a public waterway, ie. stream, creek, river or bay have a higher risk of being responsible for cleanup. If a new home owner buys a house with an oil or gas tank that has contaminated a body of water, that new owner is responsible. There are many horror stories of "new" home owners finding out that they are responsible for thousands of dollars worth of contaminated soil removal from a newly discovered leaking tank. Unfortunately, the first person who usually finds out about it after the home owner is the Realtor® . As such, both the Real Estate agent and the buyer should contractual stipulate to underground tank testing to alleviate their liability. It should be noted that in most situations home owners insurance **does not** cover the expense involved.

There are two common forms of testing available to test underground oil tanks for leakage: Soil sampling and Pressure test of tank. Soil sampling requires the pulling of soil samples from under the tank and then sending them to a laboratory where there are then analyzed for the presence of oil. Normally no more than four sample will need to be pulled unless special circumstances dictate differently. Because the samples must be sent to a lab, time should be budgeted for the lab to analyze and then report results. There are a variety of different methods for pressure testing of a tank. Normally, a vacuum test of the tank is conducted to find any leaks. Unfortunately this type of test may require that the tank have no more than 1/2" of water in bottom, no bends in the oil fill (it must be straight) and the tank must be level or the fill needs to be on the lower end of tank. Some companies are now using sensitive equipment to measure liquid loss in the tank. Either of the above test may take many hours on-site to conduct.

If a tank is found to be leaking oil or gas, the tank must either be abandoned or removed. The process of abandoning an underground tank is called *Closure.* Through this process, the tank must be pumped dry of its contents and then filled with sand or a cement slurry mix. The state of Maryland has asked those that are abandoning a tank to report it to the state so they have a record of the location of the tank. Each jurisdiction has different requirements and organizations responsible for USTs. In some areas, the only time a tank may be abandoned is if the removal "would endanger a building structure if removed." It is important that the local agency responsible for USTs be contacted to obtain local regulations. This may be the local fire department, health department, or other county agency.

The desired method for leaking USTs is removal of the tank. Following is the recommended steps when removing an underground oil tank as dictated by the Maryland Department of the Environment (MDE).

1. Notify the Oil Supplier to discontinue oil service to the home.

2. Obtain an approved contractor to conduct the work. MDE maintains a list of those approved.

3. Local jurisdictions may have specific permits to conduct removal, contact governing agency.

4. Have all oil pumped from the tank and lines. Some companies provide a credit for reusable oil removed.

5. Excavate the top of the tank and expose the piping.

6. All piping should be disconnected and drained including the tank fill line.

7. Oil sludge and residue on exterior of tank should be removed and disposed of properly.

8. If tank is clean, it can be disposed of in an approved manner. The contractor or MDE can assist in locating a proper disposal site.

9. If soil or groundwater contamination is found during excavation, it **must** be reported to MDE immediately upon discovery. Phone number: 410/631-3442 or after hours 410/974-3551.

10. Any residential tank greater than 1,100 gallons in capacity are required to be registered with MDE. Forms may be obtained by calling 410/631-3442.

Any home that has an older UST should have it periodically inspected to prevent unforeseen expense to the homeowner. Any house that is being bought and may have an older UST on site should be checked for it's condition whether it is requested by the home buyer or the Realtor® . For further information on this subject, contact your local governing agency regulating USTs, the Department of the Environment or the Environmental Protection Agency (EPA).

Carbon Monoxide

Carbon monoxide (CO) is a colorless, odorless, non-irritating gas produced as a byproduct of the incomplete combustion of fossil or wood fuels. The gas, sometimes known as "the silent killer" does not stratify, which means it can be found either high or low in a room, and generally results from improperly vented or malfunctioning combustion appliances including stoves, furnaces and hot water heaters. The gas is often entirely unnoticeable until it is too late for those who have already inhaled too much. The CO is absorbed into the bloodstream, taking the place of oxygen in the blood cells and forming Carboxyhemoglobin. This reduces the amount of oxygen available in the bloodstream, affecting all the major organs, particularly those with the highest oxygen needs, such as the brain, heart and other large muscles. Though the early signs of CO poisoning include headaches, nausea, dizziness, shortness of breath and confusion, it can quickly escalate to unconsciousness or death for the victim.

Levels

Although the effects can vary significantly based on a person's age, sex, weight and overall state of health, the following is a guideline for unacceptable levels of CO.

0 PPM - Desirable Level
9 PPM - Maximum indoor air quality level.
50 PPM - Maximum concentration for continuous exposure in any 8 hour period.
400 PPM - Frontal headaches 1 to 2 hours, life threatening after three hours.
800 PPM - Nausea and convulsions, death in two hours.
1600 PPM - Nausea within 20 minutes, death within one hour.
12,800 PPM - Death within one to three minutes.

Anyone living in a house primarily heated by oil, gas, propane, wood, or coal, is a likely candidate for CO poisoning. Over time and use, units may cease to function properly and then allow CO to enter the home. Consequently, a detector is recommended near the furnace, or any other combustion system in the home.

Some possible sources of CO in the home include:

• Unvented cooking appliances
• Wood burning fireplaces

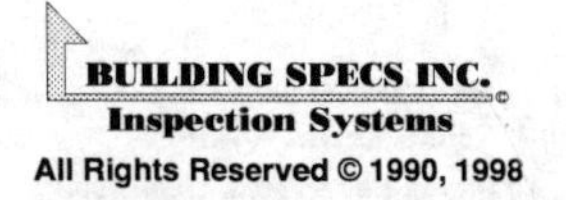

• Gas, oil, wood or coal furnaces.
• Blocked chimneys.
• Water heaters.
• Gas refrigerators.
• Gas clothes dryers.
• Attached garage (where a car may be left running).
• Barbecue grill - operated within a garage or other confined area.
• Pool/Spa heaters.
• Tobacco Smoke.
• Ceiling mounted unit heater.
• Fuel burning space heaters.

Testing

If you have concerns about CO in your home, you can purchase a CO detector to test the levels of CO in the house. There are quite a few different CO detectors on the market, and there are certain features which are recommended for these detectors. According to Consumer Reports magazine, the most accurate CO models are those of the plug-in variety with the Underwriter Labs (UL) stamp of approval. There are also sophisticated handheld units which professional CO inspectors use to test various areas around the house.

Whatever you decide to buy, try to find a constant read detector which shows the CO levels with up to the minute data. Any detector should have a Reset/Hush switch so that the buzzer indicating high CO concentrations can be turned off and the room retested. If the alarm does sound repeatedly and symptoms of CO poisoning are being felt by anyone in the house, either a qualified heating specialist or the fire department should be called to investigate further. Leave the house until they arrive to prevent further CO poisoning.

If the alarm goes off but no one feels the symptoms of CO, open windows in the house and shut off all possible sources of CO. Call a qualified repair person to investigate further.

As with all components of a house, routine inspection and maintenance should be conducted to keep systems in good working order. A chimney cleaner should be contracted for regular cleaning of flues and chimneys, and a visual inspection should also be regularly conducted. Look for rust or soot on burners, in the heat exchange, and in vents or in appliances. Also check for loose vent or chimney connections; debris or soot falling from chimneys; and moisture on the insides of windows.

Electromagnetic Fields

Electromagnetic Fields (EMF) are fields of energy generated by the use of electricity. These fields are found around all electrical power lines, electric appliances and even the electrical wiring in structures; they can also occur naturally. There is no conclusive evidence that EMF fields are detrimental to the health of people, but some researchers have noted a higher incidence of certain kinds of cancers in people who live near heavy EMF fields, like those who live near electric towers. Although every day a new conclusion is reached on the damage EMF fields may or may not cause the human body, it does not hurt to take a few simple precautions.

EMF is actually two separate energy fields - Electric and Magnetic. It is the voltage of the electricity that generates electrical fields, and the higher the voltage, the greater the electrical fields. These fields are found wherever electricity is used --- home wiring, electrical appliances and power tools --- and even though an appliance may not be on, if it is plugged in it is still generating electrical fields. These fields can be shielded by materials such as wood or metal.

Magnetic fields, on the other hand, are generated only when electricity is moving or flowing through a wire. Only when an appliance is turned on will it generate a magnetic field, and unlike electric fields, they cannot be shielded.

If possible, it is best not to spend extended periods of time near high EMF fields. For instance, if you have a choice between putting your bed directly below an electrical panel or across the room, opt for across the room. Some day we will know all the answers, but in the meantime, it is best to just play it safe.

Lead

Lead originally entered paint manufacturing in an attempt to make paint last longer. Indeed, lead proved to be excellent at improving the staying power of paint, but it soon became apparent that the hazards of lead far outweighed the benefits. As the adverse affects of lead on humans were discovered, the lead in paint was reduced; first in 1950 and then again in 1978. Consequently, the most likely homes to contain dangerous levels of lead in their paint are those built before 1950, and even if those homes have been repainted with safer paints since that time, their danger still lurks just beneath the surface.

Although lead poisoning can come from lead in water or from items like glazed pottery, the most common cause of lead poisoning is from household paint which has begun to deteriorate and turn to dust. The dust released from the deterioration is inhaled and ingested by the occupants of the house, or, sometimes children will actually chew lead covered items and absorb dangerous levels even more rapidly.

Lead is a neurotoxin, and overexposure may cause serious health problems, including injury to the nervous system, reproductive system, kidneys, blood-forming system and digestive system. Children are the most adversely affected, because children have a higher metabolism and lead can be absorbed into their systems much more quickly. Brain damage, retardation, slow mental development, irreversible learning disabilities, reduced IQ, attention deficit disorders, hyperactivity, damage to kidneys, damage to the liver and permanent neurological damage are all possible from exposure to lead dust or paint chips.

General symptoms for all sufferers include tiredness, sleeplessness, stomachache, and vomiting. It affects all of the body's systems except for the pulmonary system, where it passes straight through lungs. The greatest effects are to the central nervous system via the circulatory system, and through absorption into the bones. Because it directly affects the central nervous system, lead can cause a variety of ills including: high blood pressure, mood changes, headaches, digestive problems, nerve disorders, muscle and joint pains, kidney damage, suppressed libido, and even the lowering of sperm count.

If you note any white chalking around any painted area of your home, test the area for lead. One of the most common areas for lead paint to remain is around window panes where it was once used to fight the elements, or even on old radiators and pipes. Any one possessing an older home should make an effort to check for lead paint in these and all other painted areas. The removal of lead paint is a law for HUD homes and rental properties.

Over-the-counter lead paint tests can be purchased from hardware stores and paint stores; these contain sodium sulfide solutions which darken if lead is present; the darker the color the more lead. These test are fairly inexpensive, but their accuracy can vary due to other metals causing false positive readings as well as resins in the paint masking the lead and preventing the sodium sulfide from reacting with it. At this time, these tests have not been evaluated by the U.S. Consumer Product Safety Commission so they have not taken a position in regards to accuracy. There are two other tests that are more accurate, an X-ray Fluorescence machine and lab testing, but both require professional analysis of the sample.

According to the U.S. Department of Housing and Urban Development (HUD), if the results show a lead in pain percentage greater than 0.5% by lab testing or of greater than 1.0 milligrams per square centimeter by X-Ray Fluorescence, actions to reduce exposure should be taken. It is especially important when the paint is deteriorating or when pregnant women, infants or children are present. Depending upon the condition of the paint and economic considerations, there are different steps which can be implemented to reduce lead paint exposure in the house.

If the paint in your home contains lead you can remove the lead covered item, cover the lead paint with gypsum wallboard or spray it with a sealant, or you can have a professionals remove the dangerous paint. Although removal is the best and most permanent solution, you will probably want to vacate your home while the work is being done, for the dust caused by the removal is very toxic.

For more information:

If you have other questions or concerns about lead poisoning prevention, the following is a list of state contacts which should be able to answer you questions or provide literature.

Alabama - (205) 613-5373
Alaska - (907) 269-4940
Arizona - (602) 542-1770
Arkansas - (501) 562-7444
California - (510) 450-2453
Colorado - (303) 692-3185
Connecticut - (203) 566-5808
Delaware - (302) 739-4735
District of Columbia - (202) 767-7370
Florida - (904) 487-2945
Georgia - (404)657-6514
Hawaii - (808) 586-4254
Idaho - (208) 334-4963
Illinois - (217) 782-5830
Indiana - (317) 281-3606
Iowa - (515) 242-6340
Kansas - (913)296-1547
Kentucky - (502) 564-4830
Louisiana - (504) 765-0902
Maine - (207) 287-4311
Maryland - (410) 631-3859
Massachusetts - (800) 532-9571
Michigan - (517) 335-8246
Minnesota - (612) 627-5017
Mississippi - (601) 960-7463
Missouri - (314) 526-4911
Montana - (406) 723-0041
Nebraska - (402) 471-0197
Nevada - (702) 687-5240
New Hampshire - (603) 271-4507
New Jersey - (609) 633-2043
New Mexico - (505) 827-0006
New York - (800) 458-1158
North Carolina - (919) 733-9933
North Dakota - (701) 221-5150
Ohio - (614) 466-1450
Oklahoma - (405) 271-5220
Oregon - (503) 731-4015

Pennsylvania - (717) 783-8451
Puerto Rico - (809) 766-2817
Rhode Island - (401) 277-2808
South Carolina - (803) 737-4061
South Dakota - (605) 773-3364
Tennessee - (615) 741-5683
Texas - (512) 834-6600
Utah - (801) 538-6129
Vermont - (802) 863-7231
Virginia - (804) 371-7160
Washington - (206) 753-2556
West Virginia - (304) 558-3530
Wisconsin - (608) 266-5817
Wyoming - (307) 777-7957

Radon

Radon is a cancer-causing, radioactive gas that cannot be seen, smelled or tasted, but which causes thousands of deaths each year. Tiny radioactive particles which make up radon gas get trapped in lungs, releasing small bursts of energy as they continue to decay. These bursts can damage lung tissue and lead to cancer over time, and children are more heavily affected than adults. The Surgeon General states that radon is the second leading cause of lung cancer in the United States after cigarette smoking, and everyone is a potential victim.

Radon gas occurs from the natural breakdown of uranium in soil, rock and water. It can creep into any type of building, where it may collect until dangerous levels are reached, seriously affecting those who regularly breathe this deadly substance. It is estimated that one out of every fifteen homes in America has elevated radon levels. The only way to know whether your home or place of business is gathering radon gas is to have it tested, and if radon gas is detected, there are relatively simple steps you can take to reduce the build-up.

Radon remediation, sub slab depressurization connected to the sump-pump.

The amount of radon in the air is measured in "picocuries per liter of air" or "pCi/L, or sometimes as "Working Levels" (WL). There are many different, low cost "do it yourself" radon test kits available through mail-order or from local hardware stores, ranging from short term to long term tests. Any test you buy should display the phrase "Meets EPA Requirements," or, if you choose to hire a professional to complete radon testing, be sure they are a EPA-qualified or state certified technician.

Short Term Tests

Short term tests stay in your home for two to ninety days. Some common tests are charcoal canisters, alpha track, electret ion chamber, continuous monitors, and charcoal liquid scintillation. However, keep in mind that due to the fact that radon levels vary from day to day and season to season, short term tests are less likely than long term tests to offer you a true picture of your yearly average radon level. You could opt to use multiple short term tests over the course of the year to correct this problem.

Long Term Tests

Long term tests remain in the home for more than 90 days. Alpha track and electret detectors are commonly used for this type of testing, and should give you a good idea of your yearly average level of radon.

If, after running a radon gas test, your results are higher than 4pCi/L, you should take another short term test just to be certain. If levels are high, you will want to take steps to correct the problem before you or your family suffers health problems. Take action to reduce levels below 4pCi/L within several months if your readings are as high as 200pCi/L, and within a few weeks if higher. You should also consider testing your water if it drawn from a well. If your well water is affected, radon gas can be released into the air while you shower, increasing levels in the home. There are also some less serious affects to swallowing water contaminated by radon gas.

Common correction steps for lowering radon levels in your home include sealing cracks in floors or walls, and the installation of simple pipes and fans known as sub-slab depressurization systems. Costs for reducing the levels of radon in your home can range from $500 to $2500, but the cost to your health could be much greater if the problem is not corrected. Be sure that any contractor you might hire to correct high radon levels can and will:

1. Provide before and after test levels.
2. Explain how his repairs should lower levels.
3. Inspect your home before offering an estimate of repair costs.
4. Review the quality of your radon level results.

For more information you can call your state radon contact:

Alabama 800/582-1866
Alaska 800/478-4845
Arizona 602/255-4845
Arkansas 501/661-2301
California 800/745-7236
Colorado 800/846-3986
Connecticut 203/566-3122
Delaware 800/554-4636
District of Columbia 202/727-5728
Florida 800/543-8279
Georgia 800/745-0037
Hawaii 808/586-4700
Idaho 800/445-8647
Illinois 800/325-1245
Indiana 800/272-9723
Iowa 800/383-5992
Kansas 913/296-1560
Kentucky 502/564-3700
Louisiana 800/256-2494
Maine 800/232-0842
Maryland 800/872-3666
Massachosetts 413/586-7525
Michigan 517/335-8190
Minnesota 800/798-9050
Mississippi 800/626-7739
Missouri 800/669-7236
Montana 406/444-3671
Nebraska 800/334-9491
Nevada 702/687-5394
New Hampshire 800/852-3345 x4674
New Jersey 800/648-0394
New Mexico 505/827-4300
New York 800/458-1158
North Carolina 919/571-4141
North Dakota 701/221-5188
Ohio 800/523-4439

Oklahoma 405/271-5221
Oregon 503/731-4014
Pennsylvania 800/237-2366
Puerto Rico 809/767-3563
Rhode Island 401/277-2438
South Carolina 800/768-0362
South Dakota 605/713-3351
Tennessee 800/232-1139
Texas 512/834-6688
Utah 801/538-6734
Vermont 800/640-0601
Virginia 800/468-0138
Washington 800/323-9727
West Virginia 800/922-1255
Wisconsin 608/261-4795
Wyoming 800/458-5847

Septic Concerns

Techniques are available using magnetic tracer cables to map out your drain field.

If a house is not connected to a municipal sewer system, chances are it possess some type of septic system to treat and disperse the waste water created by bathrooms, kitchens and the laundry room. Proper maintenance of these systems is necessary to avoid polluting ground water and the high cost of having to replace theses systems.

Most septic systems possess the same general characteristics of an underground, enclosed septic tank and some sort of soil absorption system to allow the effluent to be released into the soil where it is treated by natural processes. (Cesspool will not be discussed because most jurisdictions greatly restrict or ban their use). Because the creation and operation of these systems is governed by established codes, your local health department is a very good source for information regarding your particular system, they may even be able to tell you who installed it and when.

SEPTIC TANK

All waste water from the house enters the septic tank. If the tank cannot be located, look for a sewer pipe exiting the house in the basement or crawl space. Outside of the house near the exit of the pipe is where the tank is most probably located. Some tanks have a 4 inch clean-out or inspection port marking the top of the tank. Look for plastic piping capped with a square cap, the cap usually has a small square nub on top used to wrench open the cap. Some obvious signs of placement of the septic tank are depressions in the ground, a slightly mounded patch of ground, an area on which it is difficult to grow grass or an area where snow (if applicable) melts more quickly. Septic tanks are usually rectangular in shape and can be made of concrete, fiberglass or steel. While fiberglass and concrete tanks may last up to 50 years, steel tanks may last ten years, but have been known to collapse in five years. If a property contains a steel tank, a thorough inspection of the tank should be conducted. Depending on the size of the tank and number of bedrooms in a house septic tanks can range between 1,000 to 2,000 liquid gallons. Regardless of the material the tank is made of they each have the same basic components,

The purpose of a septic tank is to separate solid (sludge) from the liquid wastes (effluent) and lighter solids (scum). Bacteria in the wastewater digest the sludge and scum and liquefy the waste products into gases and water. The gases leave via a vent, normally through the roof of the house, while the liquid is then dispersed by the drainage system and broken down by naturally occurring bacteria in the ground. The solids are collected in the bottom of the tank where it is eventually pumped by a certified septic waste removal company. To accomplish the separation between solids and liquids, the septic tank has either baffles or a tee that allows the effluent to flow out while keeping the solids trapped in the bottom of the tank.

Most newer septic tanks have an inspection or clean out pipe that will allow inspection of the tank. When looking down into the tank, the top of the tee or baffles should be visible. If they are not visible, it is possible that either they cannot be seen or they are covered over by the scum or wastewater. If this is the case, chances are the distribution pipe is clogged. This is a common problem with septic systems and a professional should be called to clear out the blockage.

EFFLUENT DISPERSAL

From the septic tank, the effluent has to be dispersed in approved manner. In almost all cases, the effluent is dispersed of into the soil where naturally occurring bacteria in the ground treat the waste. The three most common soil absorption systems are the trench system (drain or leech fields), seepage pits (dry wells), or mound systems. A distribution box maybe found if more than one trench or seepage pit are required to equally spread the effluent to each part of the system. Depending on the geographical location, any one of the three types of systems maybe in place. A standard procedure to determine which system should be in place is the Percolation (perc) test. Very simply stated, a pit is dug in the ground and a known amount of water is introduced. The ground must absorb the water in a certain period of time. If the water is absorbed within the allotted time the ground is said to perc, meaning the soil is capable of handling the liquid produced by the septic system. If the ground does not perc either another location must be chosen or a mound system may be required to disperse the effluent. What ever the circumstances, all three systems have the same purpose: to disperse the effluent in a manner that will allow the naturally occurring bacteria in the ground to break down the waste into safe substances.

Septic systems should be pumped out routinely.There are bacteria additives which are also healthy to add.

The trench system or drain field, as it is commonly known, is a perforated pipe or pipes leading from the septic tank through a layer of gravel where the effluent can be safely leeched into the ground without contaminating groundwater supplies or adjacent bodies of water. Normally, these distribution pipes are beneath one to fifteen feet of backfill, the surface of this being the yard. Between the backfill is a barrier material sitting on top of between 3/4" and 2-1/2" of gravel, with the distribution pipe running through the middle of the gravel. This system must sit at least four feet above water table or bedrock and the width of all of this is between one and three feet. (All of the above numbers vary according to local codes, call the local health department for more specific information).

Another popular type of soil dispersion system is the seepage pit or dry well. Leading from the septic tank is a distribution pipe leading into either a dug or bored well. The well is lined with either gravel or blocks with open joints. The bottom of the pit is covered with between six to twelve inches of clean gravel. Covering the well is a reinforced concrete cover with either an inspection pipe leading out of the top. (Again, all of the above numbers vary according to local codes, call your local health department for more information). Dry wells are common on properties that have small yards or yards that back to cliffs or other obstructions that prevent the use of a drain field.

The third type of soil dispersion system is the mound system. These are used almost exclusively when the ground will not perc. This may be caused by either a high water table or a rock substrate that prevents purifying the wastewater completely. Whatever the situation, the mound system is created to disperse the effluent into a man made mound that is built above ground. Leading from the septic tank, the wastewater is carried into a chamber where there is a pump that pushes it into the mound. The top of the mound is crowned with a cap (usually underneath a layer of grass and top soil). Underneath the cap is is a buffer of straw, hay, or fabric sitting on an absorption bed where the waste water enters the mound. Although different companies utilize different different methods, for the most part there is some type of fill or material that filters the effluent as it is pulled down the mound via gravity. At ground level is the earth is plowed to allow easier dispersion into the permeable soil underneath. A note about these systems: Routine inspection of the pump from the tank to the mound should be conducted. If the pump fails, the septic tank can fill and cause a backup. There should be some cover over the accessible chamber where the pump sits so that it can be inspected.

Proper use of your system and routine care and maintenance are very important in getting a long life out of the systems. Periodic pumping of the septic system every three to five years (more frequently for systems under heavy use) is necessary. Be sure that the septic company chosen pumps both the sludge and the wastewater and will dispose of it in a proper fashion. Because prices will vary , it is recommended that a couple of companies are called and prices compared. Make sure that they will also inspect the inlet and outlet baffles or tees when they come to do the pumpout, and have them repaired if there are any problems. Keep accurate records for all maintenance and repairs conducted.

HYDRAULIC FAILURE

The most common problem with septic systems is hydraulic failure. This means that the system can no longer purify the wastewater. Indications of this are strong odors emanating near the septic tank or soil absorption system, sewage and effluent coming out of the ground and ponding. Dead grass in the septic area can be another clue. Finally, if sinks and toilets do not drain properly, plumbing backs up, or gurgling sounds start occurring in the plumbing, the system may be failing. If doubts arise whether the system is operating properly, call a professional, it is the home owner's responsibility to keep the system operating properly and health departments have the ability to penalize owners if their systems fail from neglect. Leakages and contaminations can be a serious health hazard.

If the following are followed, the useful life of the septic system can be greatly extended. It is important that the septic system location is known so that it may be monitored. Heavy vehicles should stay off of the system as they can cause damage to underground piping and components. Do not build over the drainage system nor plant trees or shrubs over it, as the roots can clog drain lines. Rain water runoff from the house, spouts, sump pumps or any either water diverting devices should be kept away from the septic area to avoid overloading the system. Steps should also be taken to reduce sludge build-up in the system including; pumping of the tank, and avoiding the use of garbage disposal systems which introduce additional solids and greases which can clog the system. Garbage should be placed in the trash not down toilets or drains. This includes chemicals, paints, oils, solvents, acids, pesticides, or excessive cleaning solutions which destroy the beneficial bacteria in the tank, decrease sludge production and pollute groundwater. Lastly, take steps to conserve water: the less you use, the less entering the system.

 ***Problem:***

- Standing water in the yard.
- Depression in yard.
- Sewage Odor.
- Signs of raw sewage backing up.
- Garbage disposal.
- Flushing non-biodegradable products.
- Use of caustic chemicals or cleaners may interrupt bacteria cycle.

 ***Maintenance should include:***

- Routine Pumping.
- Adding bacteria or yeast to septic system to induce bacteria cycle.
- Clean or install a lint screen for clothes washer (lint can clog drain tile holes).

Water Contaminates

Water, no matter where its location or what its depth, is susceptible to many forms of contamination. Generally, it is required that water samples collected for water contaminate surveys (for compliance with state or county regulations) be collected by individuals approved by the state in which the test is performed.

There are many different tests which can be performed on a well, and it is best to have qualified individuals perform them to ensure the safety of those who will ultimately drink the water.

WATER TESTS

Total Coliform

This is the most common and basic test for contamination, and positive results can imply contamination from nearby or miles away.

Acceptable Levels (mg/L): 0

Possible Health Problems: Gastroenteric pathogens

Source of Contamination: Human and animal fecal waste.

Nitrate/Nitrite

Nitrate is one of the components of the nitrogen cycle. It usually occurs in low levels in surface water, and high levels may indicate biological wastes or runoff from highly fertilized fields. Groundwater sometimes naturally contains high levels of nitrates, and the level of nitrates fluctuates with seasonal changes and rain patterns.

Acceptable Levels (mg/L): 10

Possible Health Problems:
Methemoglobulinemia
Blue Baby syndrome

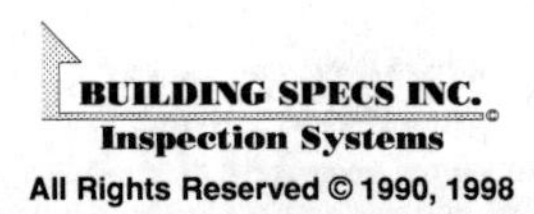

Physteria

Source of Contamination: Animal waste, fertilizer, natural deposits, septic tanks, sewage.

Lead and Copper

Lead in drinking water usually originates from soft or acidic water in the pipes eating at the faucets (chrome-plated brass) which carry it.

Any house with the following fixtures presents a potential problem.

• Lead pipes in houses built from 1910 to 1940 and even newer homes with copper piping built prior to 1986.

• Pipes soldered with lead.

• Certain submersible well pumps made with inferior brass or bronze can leach lead into the water supply. Contact the installer, pump supply house or the manufacturer to find out whether the pump contains these materials.

Acceptable Levels (mg/L): 0
Possible Health Problems: Kidney, nervous system damage.

For accurate testing the water should sit in the plumbing over night or for 8 hrs. This will allow the water to react with any lead present.

Source of Contamination:
Natural/industrial deposits, plumbing, solder, brass alloy faucets.

Acid (pH)

Low pH (under 7) indicates the water is acidic. Acidic water tends to be corrosive to metallic pipes and can lead to deterioration of the plumbing system, especially hot water heaters. In the case of copper pipes, a blue-green staining in sinks and tubs can occur. Acid water problems can generally be corrected with an acid neutralizer.

Hardness

Hardness is usually due to the presence of calcium and magnesium salts which form a curd with soaps. Most water softeners operate on an ion exchange principle, in which the well water is passed through an exchange media which is periodically regenerated with rock salt. Many water softeners increase the sodium content of water.

Chlorides

Chlorides are present in almost all natural water supplies. They can also be introduced into the water supply by seawater intrusion, ice removal chemicals and water treatment. High chloride concentrations cause objectionable tastes which are made more pronounced by softening.

Turbidity

Turbidity is a measurement of water's ability to transmit light. Ground water is commonly of

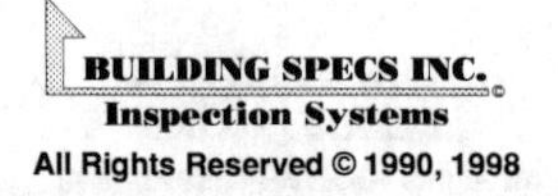

low turbidity; however, oxidized iron may elevate both color and turbidity. This may be an indicator of a low water table.

Acceptable Levels (mg/L): N/A

Possible Health Problems: Interferes with disinfection/filtration.

Source of Contamination: Soil runoff.

Iron

Iron is often the cause of reddish-brown stains in sinks and laundry, and can cause a strong metallic taste in water. Iron removal is somewhat difficult, but there are a number of "water conditioners" which may reduce the iron content to a more acceptable limit.

Radon

If a house contains a high level of radon after testing, the water source should also be tested. Because radon is a carcinogen, contaminated water needs to be filtered, or a new well in a clean area must be dug.

Acceptable Levels (mg/L): 0

Possible Health Problems: Cancer

Source of Contamination: Decay of radionuclides in natural deposits.

Because water quality is of great importance, many lending institutions require certain water test before approval of a loan. Following are examples of the protocols for water sample collections. Most lenders require that the collector and testing facility be state certified.

Collection of sample for Bacteria Test:

1. Choose sampling site. Avoid spigot with aerator or pivoting levers. Bathtub spigots are preferred. It is recommended that a little bleach solution be sprayed on the spigot of choice to kill surface bacteria.

2. Open cold water and allow to flow until it is certain that it is being pulled directly from well and not from internal plumbing. Normally ten minutes is sufficient. A decrease in water temperature is a good indication that the water is coming from the well.

3. Test for bleach in the water. Using the chemical in the aluminum package (provided by lab or testing company), pour contents into the bottom of the test tube and fill with tap water to the required line. If water turns colors (any shade of pink) the water may not be collected for testing until chlorine is run out of the system. A re-test will be necessary later.

4. If no bleach is present, fill container with water sample to recommended level and immediately put into cooler with ice (slurry) or ice pack. Sample should be taken to lab and tested

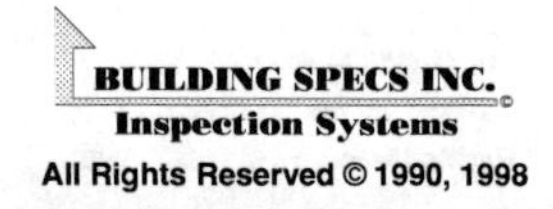

within 12 hours.

5. After testing, if no bacteria is present the water is acceptable. If bacteria is present, well disinfection is necessary.

6. After well disinfection is complete, repeat above steps for collection of a sample.

Collection of sample for Chemical Test:

1. Choose spigot, preferably from the tap that will be used as the primary drinking water tap (if bacteria test is being conducted, use the same tap.) If only conducting a chemical test, use the kitchen tap as this is the most common drinking water tap.

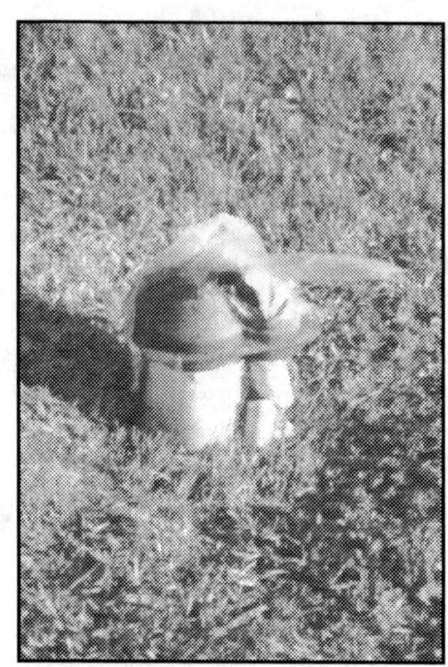

Well caps may need to be accessed if shocking becomes necessary.

2. Turn on water and conduct pH test. A pool pH kit will suffice although care should be taken that the pH kit can give a precise reading and not just a general indication of pH.

3. Collect sample as requested by laboratory, chemical samples need not be kept on ice but it is permissible to store with the bacteria sample in the cooler.

Collection of sample for Lead and Copper:

1. This test is called a first draw test. This sample should be taken before any water is used in the morning after the water has been off all evening (at least 8 hours). This way, the water has remained against the pipes as long as possible, thus creating a "worst case scenario."

Many state environmental departments offer certification courses; if a sample is going to be drawn be sure that the collector possesses the proper qualifications, or the results may not be accepted.

WELL DISINFECTION

Water obtained from private wells can be subject to bacteria contamination. If bacteria is present in the water supply, the water is not considered potable or safe for drinking. Fortunately, steps can be taken to rid the system of the bacteria.

There are various ways that bacteria can enter the water supply, including through work done on the plumbing or in the well itself. It can also enter the well through the well cap. Inspect the well cap to make sure it has a good airtight seal. Finally, in rare circumstances, the ground water itself can be contaminated, requiring filtration to eliminate the bacteria.

Because all wells are different, it is recommended that specific questions about your well be directed to a local well drilling company or your local health department. Disinfection or chlorination of the well is not a difficult procedure, but can be time consuming. Budget five to seven days to complete the disinfection. If you decide against conducting a disinfection of the well on your own, most plumbers can conduct one for you.

There are various ways that bacteria can enter the water supply. A common way for bacteria to enter the water supply is through work done on the plumbing or in the well itself. It can also enter the well through the well cap. Inspect the well cap to make sure it has a good air tight seal. Finally, in rare circumstances, the ground water itself is contaminated requiring filtration to eliminate the bacteria. Bacteria can also enter at swivel bases on sinks and collect at sink airetors.

Because all well are different, it is recommended that if you have specific questions about your well, contact a local well drilling company or your local health department. Disinfection or chlorination of the well is not a difficult procedure but it can be time consuming. Budget five to seven days to complete the disinfection. If you decide against conducting a disinfection of the well on your own, most plumbers can conduct it for you.

The two primary methods for disinfection of well utilize either regular, unscented, household bleach. Because it has very little active ingredients, it may take more than one "shocking" to completely eliminate the bacteria. As such, it is recommended that a chemical such as Sodium Dichlor or other like products be used. These are available at any swimming pool supply store.

SHOCKING PROCEDURE:

If using bleach, normally one and half to two gallons of bleach can be poured into the well head. Note: You might want to consider by-passing your water conditioner until the bleach concentration lowers to protect the plastic tank from damage. Consult the company that installed it or maintains it for more details. If using a pool chemical, one and a half to two pounds should be mixed with two to three gallons of water. (Avoid inhalation of fumes).

During this time, the chemicals are killing the bacteria in the well and plumbing lines. Following the waiting period, the chemicals must be run out of the well before another sample can be taken. *Because every well is different, we cannot tell you how long it will take to flush chemicals from well.* Turn on the outside hose and let it run (several days). Note: Do not run the water where it may get into the septic system and affect the bacteria in the tank. It is recommended water be turned off two to four hours per day to allow the aquifer to replenish.

As chemical concentration decreases, the water may be used in the house, but use your discretion. With the free chemical testing kits we provide, periodically check the chemical concentration at the outside hose. The water in the bag should not change colors. If red or pink is observed, the water is not ready to be tested. As the water turns a lighter pink, this indicates that the chemical level is decreasing. Upon obtaining a clear test outside, run the water in the house until the water does not change color inside either. As soon as the water does not change colors, call our office and we will collect a new sample.

HEALTH & ENVIRONMENTAL CHECKLIST

Health

☐ ***Further Testing and Evaluation Recommended, Which is Not Part of a Standard Home Inspection.***

1] **Radon**	☐Not Tested	☐Tested	☐Recommended________
2] **Lead Paint**	☐Not Tested	☐Tested	☐Recommended________
3] **Asbestos**	☐Not Tested	☐Tested	☐Recommended________
4] **Water**	☐Not Tested	☐Tested	☐Recommended
5] **Formaldehyde**	☐Not Tested	☐Tested	☐Recommended________
6] **Carbon Monoxide**	☐Not Tested	☐Tested	☐Recommended________
7] **Mold Spores**	☐Not Tested	☐Tested	☐Recommended________

INSPECTION SUMMARY

DATE:

OTHER NOTES (WEATHER CONDITIONS, ETC.):

PRIORITY #1 Major:

PRIORITY #2 Minor:

PRIORITY #3 Notes:

- Priority #1: *Immediate repairs, service, upgrades and/or further evaluation required by an appropriate professional.*
- Priority #2: *Potential repairs, service and/or upgrades may soon be required by an appropriate professional.*
- Priority #3: *General notes and/or annual maintenance recommendations.*

Inspection Graphic

Note: This graphic may not be to scale, and only a form of reference.

Date:

Notes:

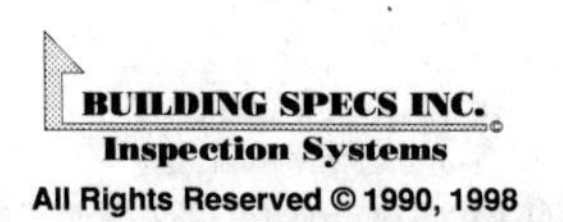

Glossary

Bridging: perpendicular framing member installed between joist and rafters at center span to help distribute load and prevent racking.

Cantilevered: Structural application. Example: floor joist overhang bearing point.

Cupping: The way a wood board curls when viewed from the end. Severe cupping can eventually cause the wood to fail or settle.

Deflection: The amount of sagging of a structural member (i.e. floor, beam). Some deflection may be pre-engineered into a beam.

Double Tapped: When two or more wires are connected to a breaker or fuse in an electric panel.

Efflorescence: Indications of water pressure (hydrostatic pressure) washing salts or minerals in the mortar, out onto the face of concrete block or brick. The presence is indicated by a white powdery substance or white staining on the surface.

Flashing: Metal or rubber connector or diverter used to divert or prevent water from entering a structure.

Gable: A style of roof framing. (A-framed).

Ground Fault Circuit Interrupter: (Abbr.: GFCI) A device used to interrupt an electrical current when an open ground, or drop in the neutral current occurs. The breaker is normally incorporated into a receptacle, but is also used in the main panel in the form of a breaker. The GFCI is identified by a test and reset button. More than one receptacle can be GFCI protected by running a series from the main GFCI device.

HVAC System: Heating-Venting-Air Conditioning System.

Ice Damming: A condition which promotes water damage at roof sheathing, where ice pushes up under roofing material due to inadequate ventilation insulation. Proper roofing membranes may prevent this from occurring.

Joist: Structural member of floor.

Moisture Meter: Tool used to observe moisture content in various materials by means of conductive or radio frequency.

NA: Abbreviation for "Not Applicable" ie: When something in the report does not apply in a certain situation.

O.C.: "On Center."

OSB: Oriented Strand Board. Pre-engineered lumber, strips of wood bound together with glue under high pressure.

NV: Abbreviation for "Not visible" ie: For the inspector to see and evaluate a situation. An invasive investigation may be required when necessary.

Piers: Independent vertical supports constructed from wood, concrete or masonry.

Punch List: Itemized list of repairs or incomplete items for new home construction.

R-Value: A rating assigned to a building product and its ability to slow heat loss, such as insulation, doors and windows.

Sheathing: Solid or spaced wood fastened to the top of the roof frame for roofing material.

Triangulation: Structural application using triangular shapes distributing compression, tension and loads used in floor and roof systems.

Warp: The way a wood board twists out of normal shape.

About The Author

Stephen L. Showalter

Stephen L. Showalter, is the founder and President of Building Specs Inc., a home inspection company made up of affiliated offices around the country. He served on the Board of Directors of the National Association of Home Inspectors (NAHI) from 1994-1998, served as President of the Maryland Association of Home Inspectors from 1996-1998, and has been a practicing home inspector since 1988. Mr. Showalter Chaired the NAHI Legislative committee, has worked on drafting proposals affecting the Home Inspection industry and testifying before the State Senate and House of Representative hearings on bills affecting the industry. He was directly involved with the formation of the Maryland Association of Home Inspectors (MAHI). Mr. Showalter has performed over 3000 fee paid inspections, conducts continuing education classes for several of the County Board of Realtors around the State of Maryland, conducts home inspection training courses, and has also been a fee-paid speaker on historic homes. He has designed and created a state-of-the-art Home Inspection Report and accompanying supplement.

With an extensive background in residential construction, Mr. Showalter has been a general contractor as well as a supervisor for the construction of multi-million dollar homes. A strong knowledge of structure, foundation, framing, roofing, and other building components has given him the background to work in many different areas of the building industry. He has over 20 years experience in custom building, and has specialized in Historic Home Renovation.